T0213961

Communications
in Computer and Information Science   1434

More information about this series at http://www.springer.com/series/7899

Arun Solanki · Sanjay Kumar Sharma ·
Sandhya Tarar · Pradeep Tomar ·
Sandeep Sharma · Anand Nayyar (Eds.)

Artificial Intelligence and Sustainable Computing for Smart City

First International Conference, AIS2C2 2021
Greater Noida, India, March 22–23, 2021
Revised Selected Papers

Springer

Editors
Arun Solanki 🆔
Gautam Buddha University
Greater Noida, India

Sandhya Tarar 🆔
Gautam Buddha University
Greater Noida, India

Sandeep Sharma 🆔
Gautam Buddha University
Greater Noida, India

Sanjay Kumar Sharma
Gautam Buddha University
Greater Noida, India

Pradeep Tomar 🆔
Gautam Buddha University
Greater Noida, India

Anand Nayyar 🆔
Duy Tan University
Da Nang, Vietnam

ISSN 1865-0929   ISSN 1865-0937   (electronic)
Communications in Computer and Information Science
ISBN 978-3-030-82321-4   ISBN 978-3-030-82322-1   (eBook)
https://doi.org/10.1007/978-3-030-82322-1

This Springer imprint is published by the registered company Springer Nature Switzerland AG
The registered company address is: Gewerbestrasse 11, 6330 Cham, Switzerland

Preface

This proceedings contains selected papers from the First International Conference on Artificial Intelligence and Sustainable Computing for Smart Cities (AIS2C2 2021). The Gautam Buddha University, India, organized this conference in collaboration with the Jan Wyzykowski University, Poland, the Petroleum-Gas University of Ploiesti, Romania, the University of Economics - Varna, Bulgaria, the WSG University, Poland, and the Federal Polytechnic, Oko, Nigeria.

AIS2C2 is focused on new technologies in the fields of AI and sustainable computing. Participants from academia and industry were invited to submit their research papers to the conference to share their ideas and information. The objective of the conference was to increase research thrust by bringing together leading scientists, academicians, researchers, students, and government officials to facilitate the exchange of thoughts and experiences. Besides this, participants were also enlightened about the vast avenues, including current and emerging technological developments, in AI and AI applications, which were thoroughly explored and discussed. Authors from countries all over the world submitted a total number of 204 papers, and the conference accepted 21 papers after a thorough peer review process, in which all papers received at least three separate reviews.

We are highly thankful to our valuable authors for their contributions and our Technical Program Committee for their immense support and motivation in making the first edition of the conference, AIS2C2 2021, a success. We are also grateful to our keynote speakers for sharing their precious work and enlightening the conference delegates. We also express our sincere gratitude to our publication partner, Springer, for believing in us.

June 2021   Arun Solanki

Organization

General Chair

Sanjay Kumar Sharma   Gautam Buddha University, India

Honorary Chairs

B. K. Panigrahi   Indian Institute of Technology, India
Mohammad S. Obaidat   King Abdullah University, UAE

Convener

Arun Solanki   Gautam Buddha University, India

Co-convener

Sandhya Tarar   Gautam Buddha University, India

Organizing Secretaries

Pradeep Tomar   Gautam Buddha University, India
Sandeep Sharma   Gautam Buddha University, India

Conference Chairs

Julian Vasilev   University of Economics, Bulgaria
Emil Pricop   Petroleum-Gas University of Ploiesti, Romania
Marzena Sobczak-Michałowska   WSG University in Bydgoszcz, Poland

Technical Program Chairs

Anand Nayyar   Duy Tan University, Vietnam
Fadi Al-Turjman   Near East University, Turkey

Conference Advisory Board

Ela Kumar   Indira Gandhi Delhi Technological University for Women, India
Chuan-Ming Liu   National Taipei University of Technology, Taiwan
Nasib Singh Gill   Maharishi Dayanand University, India
Umesh C. Pati   National Institute of Technology, Rourkela, India

Rung-Ching Chen	Chaoyang University of Technology, Taiwan
Christian Esposito	University of Salerno, Italy
Humberto Sossa	Instituto Politécnico Nacional, Mexico
Wellington Pinheiro dos Santos	Universidade Federal de Pernambuco, Brazil
Ricardo A. Ramirez-Mendoza	Tecnologico De Monterrey, Mexico
Debotosh Bhattacharjee	Jadavpur University, India
Marcos Eduardo Valle	University of Campinas, Brazil
Pompiliu Cristescu	Lucian Blaga University of Sibiu, Romania
Alessio Bottrighi	University of Eastern Piedmont, Italy
Shailendra Mishra	Majmaah University, Saudi Arabia
Mirosław Moroz	Wroclaw University of Economics and Business, Wroclaw
Aida Mustapha	Universiti Tun Hussein Onn Malaysia, Malaysia
Andriy Yerokhin	Kharkiv National University of Radio Electronics, Ukraine
Chupryna Anastasiya	Kharkiv National University of Radio Electronics, Ukraine
Abdul Wahid	MANUU, India
Alessio Bottrighi	University of Eastern Piedmont, Italy
Jagdish Chand Bansal	South Asian University, India

Publication Chairs

| Sudeep Tanwar | Nirma University, India |
| Vikrant Nain | Gautam Buddha University, India |

Publicity Chairs

Loveleen Gaur	Amity University, India
Korhan Cengiz	Trakya University, Turkey
Rudra Rameshwar	Thapar Institute of Engineering & Technology, India
P. Sanjeevikumar	Aalborg University, Denmark

Sponsor Chairs

| Sandeep Kumar | Amity University, Jaipur, India |
| G. Suseendran | Vels Institute of Science, Technology & Advanced Studies, India |

Session Coordinators

Anurag Singh Baghel	Gautam Buddha University, India
Neeta Singh	Gautam Buddha University, India
Aarti Gautam Dinker	Gautam Buddha University, India

Rajesh Mishra   Gautam Buddha University, India
Priyanka Goyal   Gautam Buddha University, India

Logistic Chairs

R. B. Singh   Gautam Buddha University, India
Vimlesh Kumar   Gautam Buddha University, India

Track Chair

Akash Tayal   Indira Gandhi Delhi Technical University for Women,
                                 India

Web Chairs

Om Prakash Sangwan   Guru Jambheswar University of Science
                                 and Technology, India
Mayank Singh   University of KwaZulu-Natal, Durban, South Africa
Hanaa Hachimi   Sultan Moulay Slimane University of Beni Mellal,
                                 Morroco

Additional Reviewers

Sunny	Manikandan	Anurag
Vikas	Bharati	Sandeep
Tanvi	Vipin	Manu
Arvind	Satish	Md Tabrez
Priyanka	Khalid	Rajalakshmi
Bhupendra	Shivani A.	Sarvesh
Ayan	Varada	Uttam
Uma N.	Nilanjan	Amanpreet
Omkar	Pushpendra	Shilpa
Sujata	Amandeep	Simar Preet
Priya	Shashikant	Vineet
Ram	Omveer	Preeti
Ashutosh	Umesh Chandra	Siddhartha
Naga	Gurpreet	Vimlesh Kumar
Aditi	Anuj	Ashish
Pradeep	Bharati	Venugopala
Amol	Vipin	Deepa
Parul	Akash	Parminder
Mamta	Shyamala Devi	Kalaiselvi

Janmenjoy

Sanjay

Mubashshir

Amit

Vinita

Amardeep Singh

Rajkumar

Veenu

Vipul

Anand

Prashant

Deepak

Naveen

Jayalakshmy

Janmenjoy

Sanjay

Niraj

Contents

Sentimental and Emotions Analysis for Smart Cities

Sentiment Analysis on the Effect of Trending Source Less News: Special Reference to the Recent Death of an Indian Actor

Aadil Gani Ganie$^{(\boxtimes)}$ ⓘ and Samad Dadvandipour ⓘ

Institute of Information Engineering, University of Miskolc, Miskolc 3515, Hungary
{ganie.aadil.gani,aitsamad}@uni-miskolc.hu

Abstract. Sentiment analysis is a developing machine learning approach to understand the text's emotions. Sentimental analysis is used to get the context of text. Virtualization of source less information is bait to further the political, religious, or personal agenda. Sentimental analysis of tweets on a late Indian actor (Sushant Singh Rajput) has been studied in this paper. Using different sentimental analysis classifiers, most of the tweets were harmful in context. Naïve Bayes proves to be the most effective classifier, with more than 82% accuracy and fewer false positive and negative ratios. Neural networks have been used with two layers on the dataset; increasing the value of epochs increases the accuracy to some level and shows an accuracy of 70.58%. Both Random Forest and SVM classifiers showed the same accuracy of 73.52% and the same false-positive and false-negative ratios.

Keywords: Sentiment analysis · Fake news · Classifiers · Hate comments

1 Introduction

Technology is bliss and dismay as well. More access to the internet leads to opinion-shaping, narrative building; sometimes it is good, like the #metoo movement, LGBTQ, etc. However, it could be damaging the social fabric of the society as well. The recent death of an actor in India (Sushant Singh Rajput) has divided society based on Twitter trends. It is a crucial source of information with more than 1/60th of the total populace utilizing it, which adds up to nearly 100 million clients A large volume of tweets are tweeted daily, and the number keeps on expanding as time passes. So to shape the narrative is straightforward as compared to the old days. It is a significantly more trustworthy source of information as the clients tweet their real sentiments and inputs, in this way, making it more reasonable for an investigation [1]. Recent tweets have been considered for the sentimental analysis and categorized into Positive, negative, and neutral. With approximately 288 million active users sharing 500 million tweets every day, tweets are among the most critically evolving datasets of user-generated content. These brief texts can convey various subjects! Opinions can help direct marketing efforts as customers share their views on brands and goods [2]. A study carried out by Alexandre Bovet [3] on US presidential elections-finds out that out of 171 million tweets, 25% of these tweets

© Springer Nature Switzerland AG 2021
A. Solanki et al. (Eds.): AIS2C2 2021, CCIS 1434, pp. 3–16, 2021.
https://doi.org/10.1007/978-3-030-82322-1_1

shared false or overly biased news. 25% is an overwhelming amount that can form a narrative in a positive or damaging manner.

While false information and scaremongering have existed since prehistoric days [4], their importance and influence are not evident in the social media era. The 2013 World Economic Forum study identified huge online misinformation as a significant techno-logical and geopolitical danger [5]; about 3.5 billion people can share 2.5 billion bytes of data from all across the world. Living in the digital ecosystem where data is power, online people can be quickly produced or misled. Users can easily communicate with everyone by clicking a button or wiping a finger and easy to get caught up in harmful or misleading social media content. The misinformation trends in online social media networks such as Facebook and other social media have recently been investigated in various studies. Facebook [6–10], YouTube [11], Twitter [9–13], or Wikipedia [14]. Aforementioned studies, as well as theoretical modeling [15–17], suggest that confirmation bias or mis-information [18] and social impact result in the proliferation in online social networks, the online social media groups that hold standard views on particular topics, i.e., echo chambers, where unsourced claims or facts, consistent with those beliefs, are likely to spread virally [8–19]. Disinformation and malicious social media posts targeting com-panies are currently on the rise. This problem hurts companies, impacting sales, stock values, brand credibility, and customer loyalty.

A recent case of online hate and disinformation is an advertisement by Tanishq jewellers (India). There was a malicious social media campaign with # BoycottTanishq on Twitter [20]. In light of the conflict, Titan Company Limited shares the brand Tanishq, dropped by 2.18% or 27.35 per share of BSE to close at ₹1,229.75 per share [21]. In general, fragmentation in cultures is often recognized in news consumption [22, 23]. It aligns with political allegiance [24]. Recent work has also exposed the function of bots, i.e., automated accounts, in spreading misinformation [25–27].

1.1 Classifiers

A large amount of data generated today is unstructured, requiring processing to derive knowledge. Examples of complex data include news stories, posts on social media, and browsing history. The method of assessing and making sense of natural language comes under Natural Language Processing (NLP). A popular NLP activity is sentiment analy-sis, which involves categorizing texts or sections of texts into a pre-defined sentiment. Sentiment analysis is a common NLP task, which entails classifying texts or parts of texts into a pre-specified sentiment. Sentiment analysis or sentiment classification falls into the wide variety of text arrangement errands where users are provided with an expression or a rundown of terms. User's classifier should tell if the sentiment behind that is fair, harmful, or impartial. In late assignments, a sentiment like "fairly sure" and "to some degree negative" is also thought (Fig. 1).

This work used three classifiers to cross-check the accuracy, false positive, and false negative ratios; the classifiers are Naive Bayes, SVM, and Random Forest. All of them show good results and accuracy of more than 80%; however, naïve Bayes proves to be most effective with an accuracy of 82%. In NLP till tokenization, all of the algorithms follow the same process. The process is:

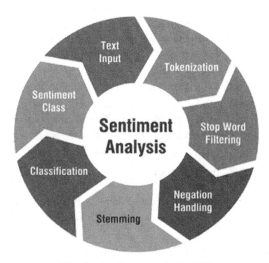

Fig. 1. Sentiment analysis [28]

- Data Gathering: Get the raw data from various sources. In our case, it is Twitter.
- Tokenize the data.
- Normalize the data.
- Stemming
- Remove noise from the data.
- Building and testing model

2 Methodology

This section has been divided into several subsections:

2.1 Data Gathering

An essential part of any analysis is based on data gathering. Tweeter data has been used for data gathering, created in 2006. Twitter is a microblogging site in which users post messages or tweets. Initially, it was planned to be a service based on SMS where messages are limited to 160 characters. Tweets are therefore limited to 140 characters, leaving a username of 20 characters. The data has been collected from hashtags using Tweepy API.

2.2 Data Cleaning

Different types of analysis require different data formats:

- **Corpus:** A collection of texts, we use pandas and Data frame for this
- **Document-term matrix**: Here, we put our data into a matrix so a machine can read it

- The term Frequency Inverse Document Frequency (TF-IDF): This is an exceptionally standard algorithm to change the text into an essential portrayal of numbers, Embeddings generated by TF-IDF are used to fit the textual data into the particular model for the forecasting and analysis.
- **Stemmer:** Stemmer maps different forms of a word into its root word. Snowball stemmer has been used for this study.
- **Lemmatizer:** Stemmers sometimes doesn't return meaningful word, especially in case of hard stemming. But Lemmatizer always returns an actual word of the language.

2.3 Exploratory Data Analysis (EDA)

It summarizes the main characteristics of the dataset. Pair plots from seaborne, word cloud, and most informative features has been used to visualize the data.

2.4 Topic Modelling

Extracting the predominant topic that people discuss the most can be done using the NLP technique called topic modeling. LDA (Latent Dirichlet Allocation) from gensim package has been used for topic modelling. The LDA approach to topic modeling is that each text is regarded to some degree as a series of topics. And each subject, again, to a certain degree, as a set of keywords. It starts with data cleaning, tokenization, then create bigrams and trigrams. A bigram is the frequency of two words occurring together, and a trigram is the frequency of three words occurring together. Number of topics choosen for the experiment are equal to 20, where each topic is a permutation of keywords, and each keyword gives a certain weightage to the topic. To understand it in a better way, we will interpret first topic 0 from the dataset. Topic 0 is represented as (0, '0.159*"family" + 0.050*"sushant" + 0.028*"meet" + 0.028*"pain" + 0.028*"search" + 0.028*"less"

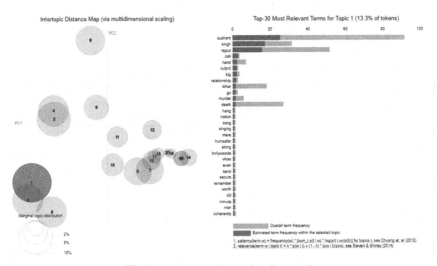

Fig. 2. Most important topics discussed

+ 0.002*"target" + 0.002*"reason" + 0.002*"singh" + 0.002*"rajput"'). The interpretation of the above statement is, the top 10 words in topic 0 are 'jail,' 'word,' 'arrest,' and so on, with jail having the highest weight. The weight reflects how important a keyword is important to that topic. After observing the weightage, In conclusion this topic carries negative sentiment and is violent in nature. Perplexity and Coherence score of topic model. Perplexity = -8.914713061826173, Coherence score = 0.4133074558694149.

Perplexity and coherence score are used to evaluate the model, lower the perplexity best the model is, higher the topic coherence, and the topic is more human interpretable (Fig. 2).

A topic is represented by each bubble on the left-hand side plot. The bigger the bubble, the more prevalent this topic is. A model with too many topics will typically have many overlaps, small-sized bubbles clustered in one region of the chart.

3   Results and Discussion

In this paper, sentimental analysis of tweets related to Indian actor Sushant Singh Rajput has been done. In India, there was a hue and a cry because of his suicide. His supporters claim that he was murdered while as a doctor's report, and the CBI team declared it a suicide. SSR committed suicide 7 Doctors Confirmed: AIIMS Forensics Head [27, 29]. Tweeter API has been used for data collection, subsequently extracted the trending tweets. To do the sentimental analysis on such tweets, tweets need to be cleaned first. Dataset was created out of these tweets and started working on them. Sentimental analysis of how people feel about death and whether they are spreading the information or misinformation has been done. For this, the tweets were divided into three categories positive, negative, and neutral. Three classifiers has been used:

- Naïve Bayes
- SVM
- Random Forest
- Neural Network

Naïve Bayes: One of the easiest supervised machine learning classifiers is the Naive Bayes Classifier. This can be trained on 90% of the data to learn what words are usually correlated with cheerful or hateful comments.

To check the distribution's shape, sorting the word counts and plotted their values on the Logarithmic axes is shown in Fig. 3. This visualization is essential to compare two or more datasets; the flatter projection suggests a broad vocabulary. In contrast, the peak distribution of restricted vocabulary is often due to a specific subject or specialized language.

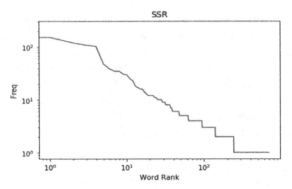

Fig. 3. Shape of distribution

Another similar plot is a histogram that shows how many words count in a given set. Of course, the distribution is highly peaked at low counts, i.e., most of the terms tend to have low counts, so we'd better show them on semilogarithmic axes to examine the tail of the distribution (Fig. 4).

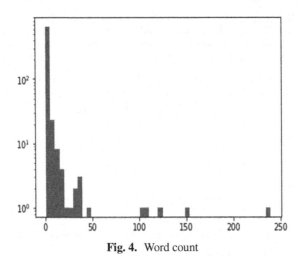

Fig. 4. Word count

Visualization of source less or fake news (Fig. 5).

Word cloud for better visualization. Word cloud is a method used to visualize recurring words in a text where words' size reflects their frequency (Fig. 6).

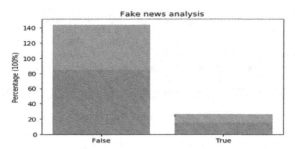

Fig. 5. Fake new analysis

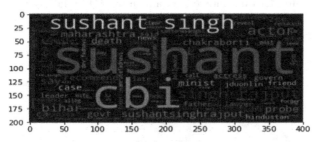

Fig. 6. Word cloud

Wordcloud for negative and positive comments in the training set, respectively (Fig. 7).

Fig. 7. Word cloud negative vs. positive comments

Accuracy is about 82.35294117647058%, which is pretty decent for such a simple model if we consider that the expected accuracy for an individual is about 80%. Finally, we can print the most insightful features, i.e., terms that often describe positive or negative comments (Fig. 8):

```
Most Informative Features
                 live = True        True : False  =     12.1 : 1.0
                  zee = True        True : False  =     12.1 : 1.0
               govern = True        True : False  =      8.7 : 1.0
                   sc = True        True : False  =      8.7 : 1.0
                minist = True       True : False  =      7.3 : 1.0
```

Fig. 8. Most informative features

3.1 Naïve Bayes Results

The confusion matrix tells us that we have a total of $(3 + 0)$ falsely labelled data out of 17 sample points. True Negative (TN), True Positive (TP), False Negative (FN), and False Positive (FP) is visualized below (Fig. 9).

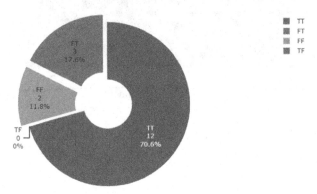

Fig. 9. Confusion matrix plot for Naïve Bayes

3.2 SVM Results

See Table 1 and Fig. 10.

Table 1. Total accuracy of 73.52%

	Prediction	Recall	F1-Score	Support
0	0.76	0.96	0.85	26
1	0.00	0.00	0.00	8
Accuracy			74.0	34
Micro Avg.	0.38	0.48	0.42	34
Weighted Avg.	0.58	0.74	0.65	34

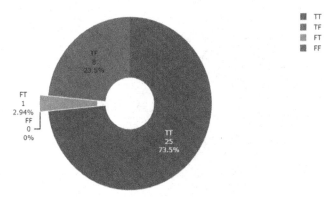

Fig. 10. Confusion matrix plot SVM

3.3 Random Forest Results

This model has been built from the training set $\{(x_i, y_i)\}\frac{n}{i=1}$ it makes predictions for its neighbours, formalized by a weighted function (Table 2 and Fig. 11),

$$y = \frac{1}{m}\sum_{j=1}^{m}\sum_{i=1}^{n}w_j(x_i, y)y_i = \sum_{i=1}^{n}\left(\frac{1}{m}\sum_{j=1}^{m}w_j(x^i, y)\right)y_i \qquad (1)$$

Table 2. Total accuracy of 73.52%

	Prediction	Recall	F1-Score	Support
0	0.76	0.96	0.85	26
1	0.00	0.00	0.00	8
Accuracy			74.0	34
Micro Avg.	0.38	0.48	0.42	34
Weighted Avg.	0.58	0.74	0.65	34

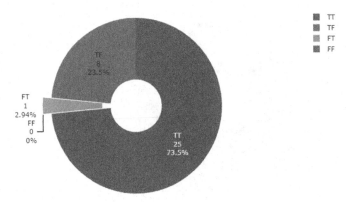

Fig. 11. Confusion matrix plot for random forest

3.4 Neural Network Results

See Table 3 and Fig. 12.

Table 3. Total accuracy of 70.58%

	Prediction	Recall	F1-Score	Support
0	0.75	0.92	0.83	26
1	0.00	0.00	0.00	8
Accuracy			71.0	34
Micro Avg.	89.9	0.46	0.41	34
Weighted Avg.	0.57	0.71	0.63	34

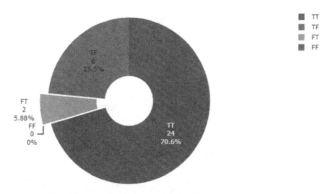

Fig. 12. Confusion matrix plot for neural networks

4 Comparison with Previous Methods

To validate our findings, comparison has been made with several existing methods. In [35] did the sentiment analysis on spam and ham emails. To verify whether they are spam using the SVM classifier and Random forest classifier, they used sentimental analysis on emails, including email content and attachment. Classifier; 96% accuracy was displayed by the former, while 97.66% accuracy was provided later. SVM showed 0% false positives and 4% false negatives, while Random Forest showed 0% false positives and 2.33% false positives (Tables 4, 5, 6, 7, 8 and 9).

Table 4. Results of [29]

Algorithm	Accuracy	Feature extraction	Dataset	Number of sentiments
SVM	45.71	Document level	IMDB	2
Naïve Bayes	65.75	Document level	IMDB	2

Table 5. Results of [30]

Algorithm	Accuracy	Feature extraction	Dataset	Number of sentiments
SVM	85.4	Uni-gram	Tweeter	2
Naïve Bayes	88.2	Uni-gram	Tweeter	2
Maximum Entropy	83.9	Uni-gram	Tweeter	2
Sematic Analysis	89.9	Uni-gram	Tweeter	2

From the above observations, it can be concluded that our model performed competitively with the existing models. With model tuning and more data, we can achieve better results. Since our model consider both regular and sarcastic text, the accuracy it achieved is competitive.

Table 6. Results of [31]

Algorithm	Accuracy	Feature extraction	Dataset	Number of sentiments
SVM	79.54	Uni-gram, bi-gram, object-oriented	Tweet	2
Naïve Bayes	79.58	Document-level	Tweet	2

Table 7. Results of [32]

Algorithm	Accuracy	Feature extraction	Dataset	Number of sentiments
SVM	90	Uni-gram	Tweet	2
Naïve Bayes	89.8	Uni-gram	Tweet	2
Maximum Entropy	90	Uni-gram	Tweet	2
Ensemble	90	Uni-gram	Tweet	2

Table 8. Results of [33]

Algorithm	Accuracy	Feature extraction	Dataset	Number of sentiments
SVM	70	Uni-gram,Sentence level	Tweet	3
Naïve Bayes	79	Uni-gram, Sentence level	Tweet	3

Table 9. Results of [34]

Algorithm	Accuracy	Feature extraction	Dataset	Number of sentiments
Naïve Bayes	85.24	Uni-gram	Tweet	2
J-48	89.73	Uni-gram	Tweet	2
BFTree	90.07	Uni-gram	Tweet	2
OneR	92.34	Uni-gram	Tweet	2

5   Conclusion

Re-tweeting and sharing unconfirmed information is harmful. We should always read and forward the information which has a good source. This paper did a sentimental analysis of Indian actors' tweets (Sushant Singh Rajput). Most of the tweets are negative, which divides society and damages the community's social fabric. Fake news or source less information has certain features like impact, scarcity, relevant topic, and virtualization. As a society, we should boycott the toxic news to prevent our next generation from falling into it. Naïve Bayes proves to be an effective classifier for doing the sentimental in categorical data with minimum false positive and false negative ratios. Naïve Bayes showed 82.35% accuracy, while as Random Forest and SVM showed the same accuracy of 73.52%. For neural networks with two layers, it showed 70.58% accuracy. Finally, concluded that it is not acceptable to trend everything on social media to suit the fake news spreaders.

References

1. Kamal, S., Dey, N., Ashour, A.S., Ripon, S., Balas, V.E., Kaysar, M.S.: FbMapping: an automated system for monitoring Facebook data. Neural Netw. World **27**(1), 27 (2017)
2. Jansen, B.J., Zhang, M., Sobel, K., Chowdury, A.: Twitter power: tweets as electronic word of mouth. J. Am. Soc. Inf. Sci. Technol. **60**(11), 2169–2188 (2009)
3. Bovet, A., Makse, H.A.: Influence of fake news in Twitter during the 2016 US presidential election. Nat. Commun. **10**(1), 1–14 (2019)
4. Soll, J.: The long and brutal history of fake news. Polit. Mag. **18**(12), 2016 (2016)
5. Howell, L.: Digital wildfires in a hyperconnected world. WEF Rep. **3**(2013), 15–94 (2013)
6. Bessi, A., Coletto, M., Davidescu, G.A., Scala, A., Caldarelli, G., Quattrociocchi, W.: Science vs conspiracy: collective narratives in the age of misinformation. PLoS ONE **10**(2), e0118093 (2015)
7. Bessi, A., et al.: Viral misinformation: the role of homophily and polarization. In: Proceedings of the 24th International Conference on World Wide Web, pp. 355–356 (2015)
8. Mocanu, D., Rossi, L., Zhang, Q., Karsai, M., Quattrociocchi, W.: Collective attention in the age of (mis) information. Comput. Hum. Behav. **51**, 1198–1204 (2015)
9. Del Vicario, M., et al.: The spreading of misinformation online. Proc. Natl. Acad. Sci. **113**(3), 554–559 (2016)
10. Del Vicario, M., Gaito, S., Quattrociocchi, W., Zignani, M., Zollo, F.: Public discourse and news consumption on online social media: a quantitative, cross-platform analysis of the Italian Referendum, arXiv Prepr. arXiv:1702.06016 (2017)
11. Shao, C., Ciampaglia, G.L., Flammini, A., Menczer, F.: Hoaxy: a platform for tracking online misinformation. In: Proceedings of the 25th International Conference Companion on World Wide Web, pp. 745–750 (2016)
12. Vosoughi, S., Roy, D., Aral, S.: The spread of true and false news online. Science **359**(6380), 1146–1151 (2018)
13. Shao, C., et al.: Anatomy of an online misinformation network. PLoS ONE **13**(4), e0196087 (2018)
14. Bessi, A., et al.: Users polarization on Facebook and Youtube. PLoS ONE **11**(8), e0159641 (2016). https://doi.org/10.1371/journal.pone.0159641
15. Ciampaglia, G.L.: Fighting fake news: a role for computational social science in the fight against digital misinformation. J. Comput. Soc. Sci. **1**(1), 147–153 (2017). https://doi.org/10.1007/s42001-017-0005-6
16. Del Vicario, M., Scala, A., Caldarelli, G., Stanley, H.E., Quattrociocchi, W.: Modeling confirmation bias and polarization. Sci. Rep. **7**, 40391 (2017)
17. Askitas, N.: Explaining opinion polarisation with opinion copulas. PLoS ONE **12**(8), e0183277 (2017)
18. Klayman, J., Ha, Y.-W.: Confirmation, disconfirmation, and information in hypothesis testing. Psychol. Rev. **94**(2), 211–228 (1987)
19. Qiu, X., Oliveira, D.F.M., Shirazi, A.S., Flammini, A., Menczer, F.: Limited individual attention and online virality of low-quality information. Nat. Hum. Behav. **1**(7), 132 (2017)

20. Tanishq withdraws advertisement on inter-faith marriage following social media criticism - The Hindu. https://www.thehindu.com/news/national/tanishq-withdraws-advertisement-on-inter-faith-marriage-following-social-media-criticism/article32841428.ece. Accessed 15 Oct 2020
21. Schmidt, A.L., et al.: Anatomy of news consumption on Facebook. Proc. Natl. Acad. Sci. **114**(12), 3035–3039 (2017)
22. Del Vicario, M., Zollo, F., Caldarelli, G., Scala, A., Quattrociocchi, W.: Mapping social dynamics on Facebook: the Brexit debate. Soc. Netw. **50**, 6–16 (2017)
23. Bakshy, E., Messing, S., Adamic, L.A.: Exposure to ideologically diverse news and opinion on Facebook. Science **348**(6239), 1130–1132 (2015)
24. Lee, K., Eoff, B., Caverlee, J.: Seven months with the devils: a long-term study of content polluters on twitter. In: Proceedings of the International AAAI Conference on Web and Social Media, vol. 5, no. 1 (2011)
25. Bessi, A., Ferrara, E.: Social bots distort the 2016 US Presidential election online discussion. First Monday **21**(11–7) (2016)
26. Ferrara, E., Varol, O., Davis, C., Menczer, F., Flammini, A.: The rise of social bots. Commun. ACM **59**(7), 96–104 (2016)
27. Sushant Singh Rajput case: CBI gets no proof of murder, now focusing on the suicide angle | Hindi Movie News - Times of India. https://timesofindia.indiatimes.com/entertainment/hindi/bollywood/news/sushant-singh-rajput-case-cbi-gets-no-proof-of-murder-now-focusing-on-the-suicide-angle/articleshow/77883136.cms. Accessed 16 Oct 2020
28. MachineX: Sentiment analysis with NLTK and Machine Learning - Knoldus Blogs. https://blog.knoldus.com/machinex-sentiment-analysis-with-nltk-and-machine-learning/. Accessed 19 Jan 2021
29. Wawre, S.V., Deshmukh, S.N.: Sentiment classification using machine learning techniques. Int. J. Sci. Res. **5**(4), 819–821 (2016)
30. Gautam, G., Yadav, D.: Sentiment analysis of twitter data using machine learning approaches and semantic analysis. In: 2014 Seventh International Conference on Contemporary Computing (IC3), pp. 437–442 (2014)
31. Le, B., Nguyen, H.: Twitter sentiment analysis using machine learning techniques. In: Thi, H.A.L., Nguyen, N.T., Van Do, T. (eds.) Advanced Computational Methods for Knowledge Engineering, pp. 279–289. Springer International Publishing, Cham (2015). https://doi.org/10.1007/978-3-319-17996-4_25
32. Neethu, M.S., Rajasree, R.: Sentiment analysis in twitter using machine learning techniques. In: 2013 Fourth International Conference on Computing, Communications and Networking Technologies (ICCCNT), pp. 1–5 (2013)
33. Hasan, A., Moin, S., Karim, A., Shamshirband, S.: Machine learning-based sentiment analysis for twitter accounts. Math. Comput. Appl. **23**(1), 11 (2018)
34. Singh, J., Singh, G., Singh, R.: Optimization of sentiment analysis using machine learning classifiers. HCIS **7**(1), 1–12 (2017)
35. Samad, D., Gani, G.A.: Analyzing and predicting spear-phishing using machine learning methods. Multidiszciplináris Tudományok **10**(4), 262–273 (2020)

Emotion Recognition Using Portable EEG Device

Aditi Sakalle[1][(✉)] [iD], Pradeep Tomar[1] [iD], Harshit Bhardwaj[1] [iD], and Arpit Bhardwaj[2] [iD]

[1] Computer Science and Engineering Department, University School of Information and Communication Technology, Gautam Buddha University, Greater Noida, India
[2] Computer Science Engineering Department, Bennett University, Greater Noida, India
arpit.bhardwaj@bennett.edu.in

Abstract. Emotions are experienced by individuals in their daily life inevitably. Emotions are integral part of man's actions separated in two distinct categories: positive and negative. There is a need to classify positive and negative emotion, as emotions play a vital role in a person's life. Therefore, in this paper, emotion classification using EEG signals with a portable EEG device is proposed. Brainwaves are recorded using a 4-channel electroencephalogram (EEG) device called MUSE2. This research evaluates and compares the performance of Multilayer Perceptron (MLP), Support Vector Machine (SVM), Genetic Programming (GP), and Hybrid Mutation based Genetic Programming (HMGP) for recognition of emotions using brainwaves. The experimental findings indicate that the proposed portable brainwaves-based emotion recognition model using the HMGP classifier provides 84.44% classification accuracy for two classes. The findings have shown better results relative to state-of-the-art approaches of the proposed system and have shown that proposed methodology will maintain greater secrecy.

Keywords: Emotion recognition · EMD · EEG · HMGP · Genetic programing

1 Introduction

Technology is an integral aspect of human activity attempts to achieve more natural contact between human and technical instruments, rising increasingly [23]. Since emotions play an essential part in human interaction, emotion identification is crucial for incorporating interaction between humans and machines and improving technology. With the rapid development of affective computing, the goal is to develop a more humanized and intelligent human-computer interaction technology for human-computer interaction (HCI) [14]. Researchers in cognitive science believe that emotions are an extensive co-factor in the cognitive process, along with speech, memory, and learning [27]. Emotions often play a critically important role in the correspondence between people, which influences the subjective interpretation of [33] text language or graphics. Relative to machines [10], the mental structure of people is more complex. If a machine can recognize human emotions, then a more efficient HCI based technology can be developed. Also, in psychology, recognizing emotions using different computational models is a well-researched area [23, 30].

© Springer Nature Switzerland AG 2021
A. Solanki et al. (Eds.): AIS2C2 2021, CCIS 1434, pp. 17–30, 2021.
https://doi.org/10.1007/978-3-030-82322-1_2

Emotion's recognition recently acquired much interest using peripheral physiologic signals [1, 2]. The physiological signal provides a better understanding of the participant's reactions generated at the experiments' time. The skin temperature [19] signals, pulse rate [28], heart rate, calculated by the Electrocardiogram (ECG) [22], are reported in this category, and by brain signals recorded using electroencephalograph (EEG) [20], functional magnetic resonance imaging (fMRI) [25], and electrocorticography (ECoG) [29]. Among these signals, the human brain generated physiological signals, which have gained more popularity in recent years. Electroencephalogram (EEG) signals are used to record the neurons' electrical activity in brains and used widely to analyze the functional changes in the brain [24]. EEG, in its association with emotions, can be considered the safest way to capture data in many different modalities due to its distinctive elements [4, 5, 8, 21].

GP approach is an essential method if the variables are continuously variable to find solutions. Since GP is a search and optimization algorithm, it can be used to develop an effective search and classification algorithm. This paper discusses the issue of emotional sensitivity by Hybrid Mutation Based Genetic Programming (HMGP). EMD is used to transform a signal into a frequency domain from the time domain. In the classification process, the aim is to improve the state-of-the-art approaches by using a new system for emotion recognition with Hybrid Mutation dependent GP (HMGP) [8]. The 10-fold cross-validation scheme illustrates proposed method's dominance and assesses the mean, minimal, and maximum precision of proposed model. The paper also shows the evaluated confusion matrix and human behavior analysis concerning the emotions' genre-wise classification.

The rest of this paper is structured accordingly. The related work is listed in Sect. 2, and the proposed method overview approach is defined in Sect. 3. Section 4 consists of experimental results, the interpretation, and the conclusion in Sect. 5.

2   Related Work

As a field of research, emotional recognition is rising. Figure 1 shows the 2-dimensional model of emotion. Emotion recognition is applied in many application areas like online gaming [16], online shopping [15], health care monitoring [6, 7, 31], and so on. Emotions profoundly affect human life as it impacts their psychological and physiological state [32]. Emotion is characterized as a stimulus-response, which lasts for seconds or minutes due to feelings [26]. Emotions are integral parts of human behavior divided into two broad categories by Davidson [12], i.e., positive emotions and negative emotions. Good emotions lead to more excellent quality and wellbeing, while negative emotions directly impact human health and reasoning. Negative feelings also lead to many mental health issues [11].

Mental disorders such as depression, tension, and anxiety stem from the buildup of negative feelings over a long time, leading to self-destruction [3]. Around 89% of India's population records mental instability, as opposed to the global average of 86% [13]. The recognition of emotion is, therefore, critical, and complicated. Therefore, a new approach needs to be established to identify effective emotions that recognize human emotions, improving human life quality. This research proposes an emotional

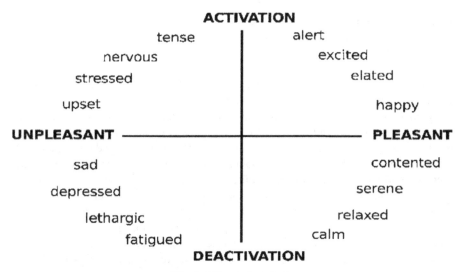

Fig. 1. 2D emotion mode

classification hybrid mutation genetic programming network for human emotions with brainwaves that are EEG signals. Robert Goodman proposed a set of 25 Questions based on 5 Scale attributes: Scale of Hyperactivity, Scale of Depressive Symptoms, Scale of Conductivity Difficulties, Scale of Peer Concerns and Scale of Proceeding, Questionnaire based emotion recognition which performed well on infants.

Li, Lan and Ji-hva Chen showed the film clip as a substance of elicitation and graded the emotion based on the Electrocardiogram (ECG), Skin Temperature, Skin Conductance (SC) and Breathing four physiological signals. Data from 60 students have been obtained. Pattern classification was carried out with a precision of 85.3%— canonical association analysis. The study revealed the potency of neuronal indications for user-dependent emotional recognition.

Basu et al. showed photographs as a material to evoke emotions in four groups. Physiological signs have been used for classifying feeling, such as Galvanic skin reaction, heart rate, breathing rate and skin temperature. The author showed that their ranking accuracy is high when the dataset size is limited, but the increase of the data accuracy declines considerably. Data from 60 students have been obtained. The study revealed the potency of neuronal indications for user-dependent emotional recognition. The classifier used was SVM, which was introduced by Wang for known emotions, using 128 EEG electrodes for three emotions. As emotional stimulation, music and videos were included.

The SVM classification obtained 6 participants and a precision of 90%. Xu et al. proposed emotional recognition using a 3-point EEG 64 electrodes; the K-Nearest Neighbor classification was used (KNN). Five participants used the picture as emotional stimuli. Bhatti et al. proposed a method to recognize emotions, i.e., happy, sad, love, and anger, in response to electronic rap, metal, rock, and hip-hop music songs. Characteristics such as time, frequency, and waves were derived from EEG signals in three distinct areas and were used to identify human emotions through the Multilayer Perceptron MLP classification. The MLP classification achieved accuracy at 78.11%. The MLP has shown better results than other states of the art methods. It was demonstrated in the paper that EEG Signals or other neuropsychological waves to predict emotions in humans had given promising results.

Various techniques for the classification of people's emotions have been pro- posed in the literature, but those approaches are restricted and resolved. The drawbacks of methods like questionnaire digital footprints, facial behavior, and others like skin temperature, etc., are manipulative because of social pressure or environmental pressure on humans. EEG based emotion recognition, on the other hand, is non-manipulative and accurate. Also, the setup of an EEG device with more no channel takes time, and the experiment can only be performed only in the technician's presence. A portable and easy to use method of emotion recognition is the requirement of the situation. The precision of the EEG- based emotion recognition strategy may decrease as the number of categorized emotions increases. The other challenge of EEG-based recognition of emotional stability is to gain signals as electrodes are located on the human scalp. The electrode carriers claim that more comfortable models of EEG devices need to be created. This work thus addresses the issue of emotional recognition while wearing a portable EEG device with emotional clips, considering two emotional classes: positive and negative, and campier different classification methods, which are MLP, SVM, GP, and HMGP. Also, the proposed HMGP classification method has shown the highest classification accuracy.

3   Method Overview

Here the definition of the dataset, feature extraction technique, Genetic Programming, and the proposed mutation operator is explained.

Dataset Description

Brainwaves are recorded using MUSE 2 EEG device to create a dataset from the pool of participants. The device is shown in Fig. 2. First, every- body from the pool was relaxed, wearing headphone Muse 2. Stimulus clips were then seen from the identified stimulus data setlist, which contains positive and negative emotions clips. The clips are arbitrarily combined in such a way that no analysis is planned. Two minutes were given to neutralize emotions—EEG- based genetic programming for emotion recognition through hybrid mutations among every clip. All the video clips have been edited in the exact resolution (720 × 576). The volume has been set to the right pace so that participants can relax with a speech. The participants were expected to sit in an enclosed room, disable both Bluetooth and wireless to avoid potential intrusion. Participants were seated roughly 0.6 m away from the middle of the screen. The brain waves are captured while the person views the stimulation clips. Thus, an EEG emotion dataset is developed as a final component. The data of the participant was collected after watching every movie clip. Participants complete a 10-point Likert scale (1 = "no" at all, 10 = "extremely") dissociation experience scale (DES) to determine the strength of each emotional self-reported. It was made up of two emotional groups. Rather than looking at the expected emotions or general mood, participants are asked to answer certain questions focusing on their current feelings. Every film clip picked produce incredibly intense emotions except for neutral videos. The order of the clips was randomized to ensure performance.

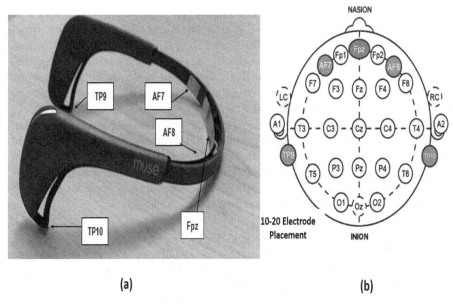

(a)   (b)

Fig. 2. Muse 2 device

Feature Extraction-Empirical Mode Decomposition

In this part the details of EEG signal's pre-processing and feature extraction technique is explained next.

Signal Pre-processing. The noise from recorded EEG signals is eliminated before the implementation of extraction methods and classification signals. The raw signals are taken with MUSE 2, and recorded EEG signals are fitted with an adjustable notch filter. An essential filter is used to exclude a particular band or interrupt a narrow bandwidth filtering. This filter is the inverse of the filter bandpass. A notch filter transmits raw EEG signals to eliminate frequencies from the 45Hz to 64 Hz band. The oversampling followed by the down sampling of signals for a sampling frequency of 256 Hz, 2uV noise (RMS) is obtained in pre-treated EEG signals from MUSE2 at the maximum sampling rate. The frontal electrode Fpz on-board driven right leg (DRL) circuit can cancel the active noise. The DRL feedback circuit also ensures proper skin contact between EEG electrodes to reduce loss. A decision tree is used to select the band in EEG signals; EEG signals' characteristics are considered input for this tree, i.e., variance, amplitude, and kurtosis. If any of these three inputs have low values, then it is regarded as a clean signal and passed on. Fast Fourier transforms the clean crude EEG signals using a 256sample window with an overlap of 90% to provide the required overlap as a method for extracting features frequency bands. MUSE 2 gives five bands named delta (1–4 Hz), theta (4–7 Hz), alpha (8- 12 Hz), beta (12–25 Hz), and gamma (25–44 Hz) bands which are passed further to extract features.

Feature Extraction-Empirical Mode Decomposition. This portion dis- cusses how the EEG signal features are recovered to recognize two emotional groups. For the extraction of features from EEG signals, EMD is used. In general, the word decomposition implies the transformation into several essential parts of a complex method or a synthesized material. In the context of signal analysis, this term is used to describe breakdowns into primary components and disrupt certain functions that did not exist when the initial data was also evolved; it covers the analysis of various sorts of sequences.

EMD is usually a way to decompose a signal without lowering the time domain to test non-linear and non-stationary signals. Any data not analytically specified and calculated on its own by the analyzed series in intrinsic mode functions (IMF) are broken down into EMD with lower-frequency oscillations in successive IMF relative to the previous sequence. At the end of this process, many bandwidth parameters would be generated, which feature the hybrid genetic programming (HMGP) classifier.

- IMF estimation with EMD on EEG signals for each iteration.
- Two properties, namely frequency parameter and amplitude parameters are calculated by the application of Hilbert transform on the IMFs for each iteration.
- Bandwidth parameters generation viz. Amplitude and frequency parameter.

Calculation of IMF. The IMF are acquired using EMD, at each iteration. In this situation, the core features are explicitly obtained from the input data.

EMD decomposes given data into intrinsic mode functions (IMF) and a residual function.

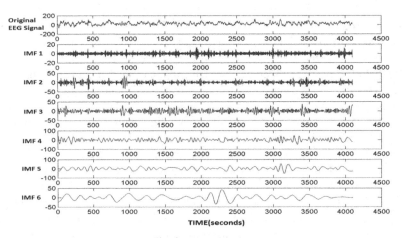

Fig. 3. EEG signals

Algorithm 1 Emotion Classification Algorithm

Input Raw EEG signals dataset.
Output Four class of emotions recognized.
Begin
Initialization Inform consent by subjects
for all watch Stimuli **do**
    EEG device setup
    while subject watch clips **do**
        EEG data acquisition
    end while end for
for all feature extraction **do**
    Apply EMD technique on
    raw EEG dataset 20 extracted
    features are f1,f2,f3,. f20
    while For classification **do**
        Load and Pass Extracted Features to MLP, SVM, GP and HMGP
        classifiers
        Classify EEG signals
    end while end for
Return: Classified two class emotions
    End

Classification

In literature, emotions are identified and categorized by various machine learning algorithms. Hybrid Genetic Software Description for human emotions with EEG signals is used in this study. The following definition is the architecture of a hybrid genetic programming model.

Hybrid Genetic Programming for Emotion Recognition. This section explains the Hybrid Mutation Genetic Programming algorithm's specifics, which are used in this research to identify the emotions. The initialization, the fitness, the crossover and hybrid mutation operators, and the termination requirements for multi-tree GPs are listed here. The following.

Initialization and Fitness Evaluation. Evolutions of tree-based structures represented by variable length expressions in a functional programming language are implementing genetic programming. Function (instructions) are the inner nodes of such framework trees. The leaves include terminals that are variable data or constant. The nodes are traversed in a fixed order (preordered or post- order) when a system tree is evaluated. By adding their feature to the results of their child nodes (subtrees), which must first be evaluated, a node's valuation is determined. And the parent node returns the value. The root node has the final function output when the execution is finished. A hybrid mutation GP classifier is initialized to classify two and four class of emotions. The size of the population is taken as 100. For two-class problem 100 single HMGP tree is created and for four class of emotion 400 HMGP trees are created as it is a multi-class classification problem i.e., $(T_1, T_2,...T_c)$. The method used to initialize the tree is Ramped Half and Half method. It is a combination of two other methods full and grow contributing 50% each. In this initialization method, half and a half ramped wide variety of tree shapes and sizes are generated, providing more exploration and exploitation space to optimize the classification problem. Instead of using the same range for the HMGP classifier, the generation processes can be defined with a "ramp". If the ramp is 3–6, there would be 25% of trees generated with a depth of 3, 25% with a depth of 5, etc. The maximum defined depth for the HMGP classifier in this framework is 6. The Function and Terminal set of this tree-based GP classifier is:

$$F = \{+, -, *, /\}$$

$$T = \{(f_1, f_2, f_3, \quad , f_{10}), R\}$$

Where the F contains the typical arithmetic operators, and the T contains 20 EEG functionalities of $(R1, f\,2, f\,3, \ldots\ldots, f\,20)$ derived from the EMD, a sum of 20 and R is an assumed floating dot count of 0.0 and 10.0. R is used to get people more diversity. R is intended to make population more dynamic. R is having value from 0.0 to 10.0 random floating-point integer.

The fitness examination is the key step in any GP strategy. The efficiency of a range of classifiers in the next new generation is calculated using this. The fitness examination is the basic test for anyone in the population. It is also easy and effective to have fitness evaluation. The fitness is measured using the formula below:

$$Fitness = \frac{TP + TN}{TP + TN + FP + FN} \tag{1}$$

where, TP (True Positive), TN (True Negative), FP (False Positive), and FN (False Negative). are used as evaluation parameters. The training of HMGP classifier is done using the 20 extracted features i.e., $F_{20} = \{f_1, f_2, \ldots f_{20}\}$ obtained from EEG signals. During training, to classify an input sample from EEG dataset i.e., F_i the classifier acts in following manner:

$$F_i(x) \geq 0 \; if \; x \in class \; i$$
$$F_i(x) < 0 \; if \; x/ \; = \; class \; i$$

It ensures that a HMGP tree classifier can categorize a x sample correctly if all samples of trees are properly categorized. On the other hand, the tree Fi, is said to correctly classify x, if Fi(x) < 0. If an input sample is classified correctly, the value of fitness function is increased by 1. Every individual is ordered in an ascending order based on their fitness after the evaluations have finished.

Crossover. Initially, the reproduction operator is implemented to the 10% of the population by passing top Nr nodes of the tree to the next generation. The genetic operator of preference typically utilizes Crossover to turn senior solutions into new, theoretically more robust approaches. So, from now on, the rest of the 30% population, i.e., Nc crossover operation, is implemented. Standard Crossover and constructive Crossover are two types of crossovers implemented to generate a new HMGP classifier generation. In standard Crossover from two parents, first, select the subtree to be swapped and develop the children; the two children are then forwarded. In constructive Crossover, the fitness of generated children is evaluated; if the child is less fit than then parent, it is discarded. This process is repeated till a fitter child than the parent is obtained. The standard Crossover gives us diversity, and constructive gives us exploitation. In this research, the classic Crossover on half of the crossover population and constructive Crossover on the other half is applied to get both diversity and exploitation in the study.

Hybrid Mutation operation. This research suggests a hybrid mutation operator to boost the GP process after a reproductive and crossover procedure and maximize the sensitivity of the HMGP cluster to detect emotions. The N_m individuals of the generation are implemented with 60% of the hybrid mutation operator. In hybrid mutation, the hill-climbing search is implemented to produce a better offspring compared to the parents of the child nodes. Here, randomly picking and substitution of a subtree with a newly generated subtree and replicate this cycle until a superior individual than the parent is reached. This hybrid mutation approach removes the negative consequence of the conversion operator by transferring the best individual to the next stage. This random generation allows us to discover a more powerful search space. It decreases the likelihood that the algorithm is stuck in the local optima. Figure 2 shows an example of a GP tree of four layers. Figure 3 shows the Hybrid Mutation operation where the fitness of generated child after mutation operation is evaluated. If the child's fitness is less then parents' fitness, it is discarded, and if the child is fitter, then parents it is accepted (Fig. 4).

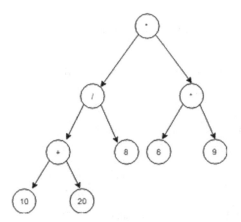

Fig. 4. GP Tree example

Termination Criteria. This HMGP approach terminates the learning/evolutionary process as it fulfills all the following requirements.

1. If the maximum number of fitness tests is reached i.e., 60,000.
2. If the classifier achieves the 100% training accuracy.

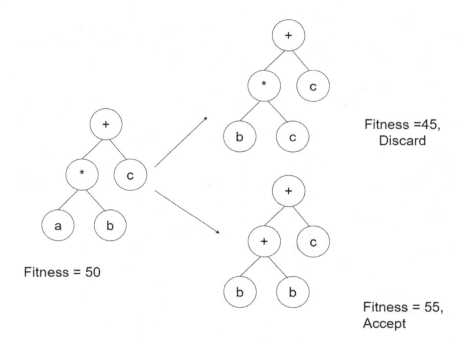

Fitness =45,
Discard

Fitness = 50

Fitness = 55,
Accept

Fig. 5. Hybrid mutation operator

The algorithm for the given process is presented as Algorithm 1 in the paper (Tables 1, 2 and 3 and Fig. 6).

Table 1. Comparison of Accuracy (%) of Proposed Work with other State-of-the-Art Methods for 2 class of emotions

Author		Classifier	Validation technique		
		Accuracy	Max.	Avg	Min
Bhatti et al. [9]	MLP	10-Fold CV	76.19	73.50	70.95
koelstra et al. [17]	SVM	10-Fold CV	81.24	78.59	73.19
Bhardwaj et al. [4]	GP	10-Fold CV	82.56	80.59	78.19
Proposed work	HMGP Classifier	10-Fold CV	84.44	83.24	81.39

Table 2. Sensitivity of Proposed Work

Author	Classifier	Validation technique	Sensitivity		
			Max.	Avg	Min
MLP	HMGP Classifier	10-Fold CV	73.44	72.24	70.39
SVM	HMGP Classifier	10-Fold CV	78.34	77.24	73.39
GP	HMGP Classifier	10-Fold CV	80.14	78.24	75.39
Proposed work	HMGP Classifier	10-Fold CV	83.44	82.24	80.39

Table 3. Specificity of Proposed Work

Author	Classifier	Validation technique	Specificity		
			Max.	Avg	Min
MLP	HMGP Classifier	10-Fold CV	71.44	69.24	65.39
SVM	HMGP Classifier	10-Fold CV	73.34	72.24	70.39
GP	HMGP Classifier	10-Fold CV	77.45	75.14	73.39
Proposed work	HMGP Classifier	10-Fold CV	85.24	83.64	82.12

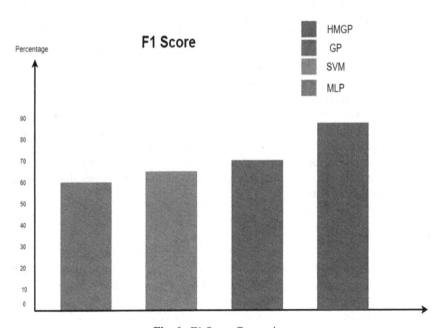

Fig. 6. F1 Score Comparison

4   Experimental Results

This section demonstrates the experimental findings of the suggested Hybrid Mutation based Genetic System for Emotional Recognition by EEG Signal Analysis, coupled with three separate state-of-the-art MLP [9], SVM [17], GP [1] classification models. There are two types of positive and negative emotions included in each of these algorithms. In Python (3.6) and Intel, i7 7th gen laptop with 16 GB of RAM, the GP-based HMGP, MLP, SVM, and GP is implemented. Of all specific classifiers, the extraction method is EMD. The parameters for applying MLP, SVM and GP are the same. as [9, 17]. Initial GP experiments with the regular crossover operator as defined by Koza [18] are used to evaluate the significance of the parameters primarily based on heuristic guidance on the choice of parameters and methodological quest. The experimental results are presented in two stages: (1) human emotion recognition performance evaluation using accuracy and confusion matrix and (2) human behavior analysis in terms of responsiveness for recognizing emotions by different age groups. In this, the entire dataset is partitioned into approximately ten equal size blocks. 90% of the dataset, i.e., nine blocks, becomes the training data and 10% of the dataset, i.e., one block, becomes the testing data. This process continues ten times, such that a new data block is used for testing every time. Using 10-fold cross-validation, the precision and confusion matrix in all types in the two and four emotions categories are examined. The EMD method is used to extract ten attributes from the brain wave dataset EEG rather than interpret individual emotions by utilizing four separate classifiers. The results show that the accuracy of the proposed work with other State-of-the-Art Methods for 2 class of emotions is high in comparison. The maximum accuracy of MLP, SVM, GP and proposed HMGP is 76.19, 81.24, 82.56, and 84.44, respectively, the average accuracy of MLP, SVM, GP and proposed HMGP is 73.50, 78.59, 80.59, and 83.24, respectively. The minimum accuracy of MLP, SVM, GP and proposed HMGP is 70.95, 73.19, 78.19, and 81.39, respective. The sensitivity and specificity of proposed model to describe the results in detail is evaluated. The maximum values for Sensitivity and Specificity for the proposed classifier HMGP are 83.44 and 85.24, respectively, confirming proposed model's effectiveness. The F1 score is also evaluated to demonstrate that proposed model works efficiently on unbalanced data. Figure 5 compares the F1 score with other methods like MLP, SVM, and GP. It has been cleared from the results that this proposed method outperforms the other way of handling unbalanced data.

5   Conclusion

Hybrid mutation-based GP is successfully applied to the recognition of emotion task in this research work. In terms of specification and network architecture, the architecture of this HMGP network studies changed considerably. The two separate modules are proposed: stimuli clips, EEG data gathering, extraction of features and emotion classification. For this reason, a hybrid mutation-based genetic programming classifier, which has an excellent accuracy compared with the state-of-the-art approaches, is defined by EMD after feature extraction. Many trials have helped contrast HMGP methodology to current practices to verify these findings. Such findings reveal that some related discreet emotions are reliable and capable of analyzing the existing state-of-the-art approaches. This program uses a commercially accessible EEG-reading tool, Muse 2, for handheld EEG.

References

1. Acharya, D., Billimoria, A., Srivastava, N., Goel, S., Bhardwaj, A.: Emotion recognition using Fourier transform and genetic programming. Appl. Acoust. **164**, 107260 (2020)
2. Acharya, D., Goel, S., Asthana, R., Bhardwaj, A.: A novel fitness function in genetic programming to handle unbalanced emotion recognition data. Pattern Recogn. Lett. **133**, 272–279 (2020)
3. Al-Shargie, F., Kiguchi, M., Badruddin, N., Dass, S.C., Hani, A.F.M., Tang, T.B.: Mental stress assessment using simultaneous measurement of EEG and fNIRS. Biomed. Opt. Express **7**(10), 3882–3898 (2016)
4. Bhardwaj, A., Tiwari, A., Vishaal Varma, M., Ramesh Krishna, M.: Classification of EEG signals using a novel genetic programming approach. In: Proceedings of the Companion Publication of the 2014 Annual Conference on Genetic and Evolutionary Computation, pp. 1297–1304 (2014)
5. Bhardwaj, A., Tiwari, A., Vishaal Varma, M., Ramesh Krishna, M.: An analysis of integration of hill climbing in crossover and mutation operation for EEG signal classification. In: Proceedings of the 2015 Annual Conference on Genetic and Evolutionary Computation, pp. 209–216. ACM (2015)
6. Bhardwaj, A., Tiwari, A., Bhardwaj, H., Bhardwaj, A.: A genetically optimized neural network model for multi-class classification. Expert Syst. Appl. **60**, 211–221 (2016)
7. Bhardwaj, A., Tiwari, A., Krishna, R., Varma, V.: A novel genetic programming approach for epileptic seizure detection. Comput. Methods Programs Biomed. **124**, 2–18 (2016)
8. Bhardwaj, H., Sakalle, A., Bhardwaj, A., Tiwari, A.: Classification of electroencephalogram signal for the detection of epilepsy using innovative genetic programming. Expert Syst. **36**(1), e12338 (2019)
9. Bhatti, A.M., Majid, M., Anwar, S.M., Khan, B.: Human emotion recognition and analysis in response to audio music using brain signals. Comput. Hum. Behav. **65**, 267–275 (2016)
10. Chen, M., Zhang, Y., Qiu, M., Guizani, N., Hao, Y.: SPHA: smart personal health advisor based on deep analytics. IEEE Commun. Mag. **56**(3), 164–169 (2018)
11. Clifford, G., Hitchcock, C., Dalgleish, T.: Negative and positive emotional complexity in the autobiographical representations of sexual trauma survivors. Behav. Res. Ther. **126**, 103551 (2020)
12. Davidson, R.J.: On emotion, mood, and related affective constructs. The nature of emotion: Fundamental questions, pp. 51–55 (1994)
13. Deb, S., Strodl, E., Sun, J.: Academic stress, parental pressure, anxiety and mental health among Indian high school students
14. Hossain, M.S.: Patient state recognition system for healthcare using speech and facial expressions. J. Med. Syst. **40**(12), 1–8 (2016). https://doi.org/10.1007/s10916-016-0627-x
15. Shamim Hossain, M., Muhammad, G.: An emotion recognition sys- tem for mobile applications. IEEE Access **5**, 2281–2287 (2017)
16. Kim, Y., Lee, H., Provost, E.M.: Deep learning for robust feature generation in audiovisual emotion recognition. In: 2013 IEEE International Conference on Acoustics, Speech and Signal Processing, pp. 3687–3691. IEEE (2013)

17. Koelstra, S., et al.: Single trial classification of eeg and peripheral physiological signals for recognition of emotions induced by music videos. In: Yao, Y., Sun, R., Poggio, T., Liu, J., Zhong, N., Huang, J. (eds.) BI 2010. LNCS (LNAI), vol. 6334, pp. 89–100. Springer, Heidelberg (2010). https://doi.org/10.1007/978-3-642-15314-3_9

18. Koza, J.R.G.P.: On the programming of computers by means of natural selection. Genet. Program. (1992)

19. Liapis, A., Katsanos, C., Sotiropoulos, D., Xenos, M., Karousos, N.: Recognizing emotions in human computer interaction: studying stress using skin conductance. In: Abascal, J., Barbosa, S., Fetter, M., Gross, T., Palanque, P., Winckler, M. (eds.) INTERACT 2015. LNCS, vol. 9296, pp. 255–262. Springer, Cham (2015). https://doi.org/10.1007/978-3-319-22701-6_18

20. Lin, Y.-P., et al.: Eeg-based emotion recognition in music listening. IEEE Trans. Biomed. Eng. **57**(7), 1798–1806 (2010)

21. Lokannavar, S., Lahane, P., Gangurde, A., Chidre, P.: Emotion recognition using EEG signals. Emotion **4**(5), 54–56 (2015)

22. Marín-Morales, J., et al.: Affective computing in virtual reality: emotion recognition from brain and heartbeat dynamics using wearable sensors. Sci. Rep. **8**(1), 13657 (2018)

23. Meng, X., Wang, S., Liu, H., Zhang, Y.: Exploiting emotion on reviews for recommender systems. In: Thirty-Second AAAI Conference on Artificial Intelligence (2018)

24. Novák, D., Lhotská, L., Eck, V., Sorf, M.: EEG and VEP signal processing. Cybernetics, Faculty of Electrical Engineering, pp. 50–53 (2004)

25. Luan Phan, K., Wager, T., Taylor, S.F., Liberzon, I.: Functional neuroanatomy of emotion: a meta-analysis of emotion activation studies in pet and fMRI. Neuroimage **16**(2), 331–348 (2002)

26. Picard, R.W., Vyzas, E., Healey, J.: Toward machine emotional intelligence: Analysis of affective physiological state. IEEE Trans. Pattern Anal. Mach. Intell. **10**, 1175–1191 (2001)

27. Qing, C., Qiao, R., Xiangmin, X., Cheng, Y.: Interpretable emotion recognition using EEG signals. IEEE Access **7**, 94160–94170 (2019)

28. Ragot, M., Martin, N., Em, S., Pallamin, N., Diverrez, J.-M.: Emotion recognition using physiological signals: laboratory vs. wearable sensors. In: Ahram, T., Falcão, C. (eds.) AHFE 2017. AISC, vol. 608, pp. 15–22. Springer, Cham (2018). https://doi.org/10.1007/978-3-319-60639-2_2

29. Paramesura Rao, V.R., Hewawasam Puwakpitiyage, C.A., Muhammad Azizi, M.S.A., Tee, W.J., Murugesan, R.K., Hamzah, M.D.: Emotion recognition in e-commerce activities using EEG-based brain computer interface. In: 2018 Fourth International Conference on Advances in Computing, Communica tion & Automation (ICACCA), pp. 1–5. IEEE (2018)

30. Russell, J.A.: Measures of emotion. In: The Measurement of Emotions, pp. 83–111. Elsevier (1989)

31. San-Segundo, R., Gil-Martín, M., D'Haro-Enríquez, L.F., Pardo, J.M.: Classification of epileptic EEG recordings using signal transforms and convolutional neural networks. Comput. Biol. Med. **109**, 148–158 (2019)

32. Sourina, O., Wang, Q., Liu, Y., Nguyen, M.K.: A real-time fractal-based brain state recognition from EEG and its applications. In: Biosignals, pp. 82–90 (2011)

33. Wagner, J., Kim, J., Andr´e, E.: From physiological signals to emotions: implementing and comparing selected methods for feature extraction and classification. In: 2005 IEEE International Conference on Multimedia and Expo, pp. 940–943. IEEE (2005)

Classification of Extraversion and Introversion Personality Trait Using Electroencephalogram Signals

Harshit Bhardwaj[1]([⊠]) ⓘ, Pradeep Tomar[1] ⓘ, Aditi Sakalle[1] ⓘ, and Arpit Bhardwaj[2] ⓘ

[1] Department of Computer Science and Engineering, University School of Information and Communication Technology, Gautam Buddha University, Gautam Budh Nagar, Greater Noida, Uttar Pradesh 201312, India
[2] Department of Computer Science Engineering, Bennett University, Gautam Budh Nagar, Greater Noida, Uttar Pradesh 201312, India

Abstract. Conventional methods for assessing personality include individual feedback questionnaires and personality assessment through social networking platforms such as Facebook, Twitter, image Instagram, film, and an online sentiment study. In response to specific trait video clips, this work offers a framework for identifying personality traits in real-time. These film clips create sensations in human beings, and we captured the brain signals with the electroencephalogram (EEG) system and analyze their personality traits throughout that time. This work considered the personality trait of Extraversion and Introversion for personality prediction. This method relies on Fast Fourier Transform (FFT) to extract features and the Genetic Programming (GP) to classify EEG details. Experiments have been conducted using EEG data collected from a single NeuroSky Mind Wave 2 dry electrode device. Such findings have shown enhancement in the state of art methods and have verified the possible use of our approach to predict these traits.

Keywords: Personality prediction · Electroencephalogram signals · Fast Fourier Transform · Genetic programming

1 Introduction

Personality [1] is a blend of all the character and qualities that make a person's personality unique. Psychology assumes that individuals vary in their outlook according to several fundamental aspects that persist over time and in different situations. Personality characteristics represent people's thinking, feeling, and behavioural patterns [2]. The most widely used trait system is the Five-Factor Model (FFM) [3] and the Myers Briggs Type Indicator (MBTI) model [4]. The FFM has five personality traits, whereas the MBTI has four areas of traits. Out of four areas, one area consists of the Extraversion and the Introversion trait.

Personality prediction is mainly made using social media data like Facebook, Twitter, etc. [5]. Some authors have done personality prediction using a set of questionnaires [6]. Personality is also predicted using physiological signals [7]. But the personality

© Springer Nature Switzerland AG 2021
A. Solanki et al. (Eds.): AIS2C2 2021, CCIS 1434, pp. 31–39, 2021.
https://doi.org/10.1007/978-3-030-82322-1_3

prediction using the EEG signals has gained the higher classification accuracy, and the most important thing about the EEG signals is that no one can fake them; thus, we get accurate results [8].

This research aims to build an automated personality prediction classifier that predicts that the user has either Extraversion or Introversion traits.

Further paper is presented in the following manner: Sect. 2 discusses the related work done for personality prediction. Section 3 includes the methodology to construct the classifier for personality prediction, and Sect. 4 address the experimental results.

2   Related Work

Kartelj et al. [9] proposed a novel approach Automated Personality Classification (APC), using different combinations of the APC corpora, psychological trait measurements, and learning algorithms for social networks. Next, they explored possible enhancements to the existing APC approach, which uses various APC corpora combinations, measures of psychological features, and learning algorithms. The author also subsequently considers extensions to the APC question and related tasks, such as complex APC and personality differences in a file. All this study was conducted in connection with social networks and data processing processes associated with them.

Fisher et al. [10] present various hypotheses around the idea of humans having evolved to a common fundamental personality trait and offering evolutionary advantages. Based on the emerging GFP theory, socio-analytical theory, and latest advances in personality research, the researcher utilized this method to analyze the amount and favourable association between homogenous subtypes and proximal indicators of evolutionary health, including existence and job satisfaction, work motivation, and moral values. This study creates three latent profiles, which are defined by the degree of awareness, extraversion, cooperation, and emotional maturity that participants embrace. The characteristic of openness towards experience does not appear to 7 distinguish between the three latent types.

Pratama et al. [11] research, the user's personality was predicted successfully on the Twitter text. Social media users can review posts to collect their personally identifiable information. This experiment predicts personality based on text written by text-classified Twitter users. English and Indonesian are the languages used. Naive Bayes, K-Nearest Neighbours, and Support Vector Machine are classification methods implemented. The results of the experiments showed that the other strategies were marginally better than Naive Bayes. Naive Bayes has made the other approaches of 60% marginally stronger by three processes. Experimental accuracy from previous studies does not increase. Compared with checking by questionnaires, the device has 65% accuracy.

Adi et al. [12] introduce Twitter-based optimization techniques for Automatic Personality Recognition (APR) in Indonesian Bahasa. This work discusses Bahasa Indonesia's optimization of APR. To advance the Machine Learning algorithms, the author tests a set of techniques for hyperparameters tuning, function selection, and sampling. The prediction method is based on the algorithms of machine learning. The study uses three Machine Learning algorithms: Stochastic Gradient Descent and two learning algorithms, Gradient Boosting and stacking (super learner). Additional development may be accomplished with the expansion of the data set and the use of deep learning algorithms to predict user personality automatically.

Harrington et al. [13] research included the MBTI model and a Likert-style questionnaire, mainly 166 female students, two-thirds taking or having studied four or more online courses, who questioned why they had chosen to offer a teaching form. Tests showed that a large proportion of Introverts prefers face-to-face online classes and Extroverts. The current research showed that MBTI students preferred online classes to a statistical majority Introvert type, while students preferred face-to-face classes to Extravert type. There are other limits on the new one. The key limitation is that the gender sample was not balanced.

Shen et al. [14] discuss how Facebook demonstrates neurotic and extraversion characteristics in a user's behaviour. The author developed in the previous study a Facebook application to directly recover the data of 1327 users based on self-reported Facebook use by college students. A total of 154 attributes were assessed, covering ages, details of personalities, generated content, and relationships with friends. Patients with neurotic disorders tend to disclose their knowledge. Extroverts' friends on Twitter are far more than introverts. Twitter is less likely to insulate neurotics and introverts. It helps us analyze and compare user-specific actions such as writing styles or "likes" with fine signals. We present significant features and prove that they can deduce the neuroticism and extraversion of a person.

3 Methodology

This section presents the detailed methodology of implementing the classifier for personality prediction and includes the details of the dataset, classification algorithms used, the procedure for conducting the experiment, and GP life cycle [15–19] using constructive crossover are provided in this section.

3.1 Dataset

The dataset contains EEG signals [20–23] of 15 participants using the NeuroSky Mind Wave Mobile 2 device. The movie clips have been taken from the IMDB database [24]. The dataset also contains questionnaires that participants must fill after watching a video clip targeting extraversion and introversion trait.

3.2 Feature Extraction Technique

After dataset creation, we apply the Fast Fourier transform [25] feature extraction technique on the dataset to extract the important features. The result of this step is that ten features are extracted as output. These ten features become the input for our classifier.

A fast Fourier transform (FFT) is a technique that calculates the discrete Fourier transform of an IDFT sequence. The Fourier method converts a signal from the source domain (often space or time) into the frequency domain, vice versa. The FFT is accomplished by splitting a sequence of values into different frequency components.

3.3 Classification Algorithms

Numerous methods for personality prediction are included in the literature. The KNN and GP algorithms are used in this work for the classification of personality traits with EEG signals. The description of the classifiers is given below.

3.3.1 KNN

K-Nearest Neighbour (KNN) [11] is the easiest algorithm based on supervised learning methods for Machine Learning. KNN is used for statistical problems both in classification and for regression. However, it is mainly used in the industry for statistical classification issues. The Algorithm for KNN uses 'change' to predict new datapoints values.

3.3.2 Random Forest

Random Forest (RF) [26] is a common mechanism for machine learning that belongs to the supervised learning technique is Random Forest. For classification as well as regression criteria at ML, it can be used. It is based on the concept of ensemble learning. On different subsets of the given data set, RF is a classifier containing several decision trees and takes the median to boost that dataset's precise quality. Instead of focusing on one decision tree, the random forest takes prediction from the individual tree and based on multi vote estimates and predicts the results.

3.3.3 Preliminaries of GP

We had to follow the following steps to build the GP-based classifier [27, 28], also known as Constructive Genetic Programming (CGP) classifier for EEG classification. The parameter values of GP are taken from the paper [29].

- **Initialization**
  We will first produce the initial population for GP to carry out operations. The terminal set (TS) and the function set (FS) are essential components to build the initial population. The TS and FS are the system alphabets to be generated. The TS collection is made up of system variables and constants. The FS are multiple math functions, including adding, subtracting, and dividing, multiplying, etc. The FS consists of arithmetic and TS, with feature variables and constants, randomly initialised for increasing of the trees [30]. The FS and TS used in this section are:
  FS = {+, −, *, %} and.
  TS = {feature variables, R}.
  Where R includes [0.0 to 10.0] constants randomly generated. We have initialized trees using the ramped half and a half approach [31].
- **Fitness Measure**
  Fitness measure is one of GP's most important concepts [32–34]. Fitness Measure is part of GP's life cycle. Also, by using the fitness function can the performance of the system be calculated. The fitness function lets GP looking for a computer program that is most capable for a specific mission.

$$\text{Fitness} = \frac{\text{TP} + \text{TN}}{\text{TP} + \text{TN} + \text{FP} + \text{FN}}$$

TP = True Positive
TN = True Negative
FP = False Positive
FN = False Negative

- **Genetic Operators**
  Once the fitness value of all the individuals generated during the initial population is calculated, those individuals' genetic operators are applied to improve their fitness.
- **Reproduction**
  The top N_R (Probability of Reproduction) individuals with the highest fitness value transferred directly to the next generation.
- **Constructive Crossover**
  A local hill-climbing technique (LHC) is blended into the crossover operation during the constructive crossover (CC) [35, 36]. After the reproduction operation, the rest of the Nc individuals perform the CC operation. In CC, individuals become pairs and undergo a crossover process. The offspring produced by chosen pairs are matched with the parent's fitness. When contrasted, they are approved only when the offspring's fitness is better than their parents. They are denied otherwise, and the cycle is replicated until we get better fitness offspring.
- **Constructive Mutation**
  In the constructive mutation (CM), the individuals are chosen by applying an LHC for generating the offspring better than their parents. We pick a subtree randomly from the parent and then substitute it using a newly created subtree and repeat this process until better offspring in terms of fitness is generated.
  The goal of the CC and CM operation is to improve the fitness of offspring and the classification accuracy.
- **Termination Criteria**

The evolution process ends when one of the conditions is fulfilled:

i. The number of generations exceeds the maximum count.
ii. In the training set, the classification problem was solved, i.e., all items of interest in the training set were correctly classified [36].

4   Experimental Results

This section gives details about the KNN and GP classifier's accuracy results designed for the personality prediction. Our dataset is divided into 50–50 and 60–40 partition schemes, and the total samples are 1850. So, the training set for the 50–50 partition scheme contains 925 samples, and the testing set contains 925 samples. So, the training set for the 60–40 partition scheme contains 1110 samples, and the testing set contains 740 samples. Table 1 represents the comparison of the classification accuracy of the implemented classifiers. Figure 1 represents the classification accuracy of the implemented classifiers in the 50–50 partition scheme. Figure 2 represents the comparison of classification accuracy of the implemented classifiers in the 60–40 partition scheme. Table 2 represents the comparison of the sensitivity, precision, and specificity of the KNN, RF and CGP classifiers.

Table 1. Comparison of classification accuracies of classifiers

Classifier	Partition scheme	Accuracy		
		Minimum	Average	Maximum
KNN	50–50	57.63	60.45	63.64
	60–40	60.14	62.79	65.07
RF	50–50	63.38	66.73	68.57
	60–40	65.08	67.29	70.43
CGP	50–50	66.93	69.97	71.86
	60–40	68.43	71.15	73.54

The findings reveal that the CGP classifier defeats the accuracy of the KNN and RF classifier. The rise in the CGP classifier accuracy is because of the constructive crossover operator. The minimum, average, and maximum classification accuracy of the CGP classifier is 68.43%, 71.15%, 73.54%, respectively.

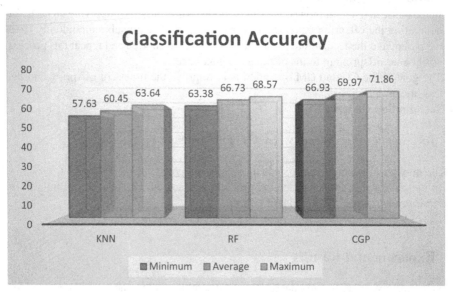

Fig. 1. Comparison of classification accuracy of KNN, RF, and CGP classifier in 50–50 partition scheme

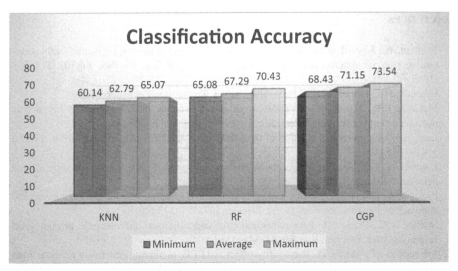

Fig. 2. Comparison of classification accuracy of KNN, RF, and CGP classifier in 60–40 partition scheme

Table 2. Comparison of sensitivity, precision and specificity of KNN, RF and CGP

Classifier	Partition scheme	Sensitivity	Precision	Specificity
KNN	50–50	62.47	61.35	60.14
	60–40	64.87	63.84	61.89
RF	50–50	67.78	66.76	65.07
	60–40	69.48	68.73	66.53
CGP	50–50	71.65	70.72	68.54
	60–40	73.45	72.76	70.78

5 Conclusions

In this study, a framework for identifying personality traits in real-time, for which two classifiers are implemented, also known as KNN and CGP classifier for the EEG based personality prediction, which classifies that the person is having either introversion trait or extraversion trait. The FFT technique is applied for feature extraction. The results demonstrate that the CGP classifier accuracy is higher than the KNN classifier in classifying personality traits. In the future, we will try to apply more machine learning algorithms to our dataset and try to enhance the classification accuracy.

References

1. Norman, W.: Toward an adequate taxonomy of personality attributes: replicated factor structure in peer nomination personality ratings. J. Abnorm. Soc. Psychol. **66**(10), 574–583 (1963)
2. Eysenck, H.J.: Dimensions of Personality, vol. 5. Transaction Publishers (1950)
3. McCrae, R.R., Costa, P.T.: Validation of the five-factor model of personality across instruments and observers. J. Pers. Soc. Psychol. **52**(1), 81 (1987)
4. Lorr, M.: An empirical evaluation of the MBTI typology. Personal. Individ. Differ. **12**(11), 1141–1145 (1991)
5. L¨onnqvist, J-.E., Deters, F.: Facebook friends, subjective well-being, social support, and personality. Comput. Hum. Behav. **55**, 113–120 (2016)
6. Pilarska, A.: Big-five personality and aspects of the self-concept: variable and person centered approaches. Personal. Individ. Differ. Elsevier **127**, 107–113 (2018)
7. Stemmler, G., Wacker, J.: Personality, emotion, and individual differences in physiological responses. Biol. Psychol. Elsevier **84**, 541–551 (2010)
8. Schmidtke, J.I., Wendy, H.: Personality, affect and EEG: predicting patterns of regional brain activity related to extraversion and neuroticism. Personal. Individ. Differ. **36**, 717–732. (2004) ISSN 0191–8869
9. Kartelj, A., Vladimir, F., Milutinović, V.: Novel approaches to automated personality classification: Ideas and their potentials. In: 2012 Proceedings of the 35th International Convention MI- PRO, IEEE (2012)
10. Fisher, P.A., Robie, C.: A latent profile analysis of the five-factor model of personality: a constructive replication and extension. Personal. Individ. Differ. Elsevier **139**, 343–348 (2019)
11. Pratama, B.Y., Sarno, R.: Personality classification based on twitter text using Naive Bayes, KNN and SVM. In: 2015 International Conference on Data and Software Engineering (ICoDSE), pp. 170–174. IEEE (2015)
12. Adi, G.Y.N.N., et al.: Optimization for automatic personality recognition on Twitter in Bahasa Indonesia. Proc. Comput. Sci. **135**, 473–480 (2018)
13. Harrington, R., Loffredo, D.A.: MBTI personality type other factors that relate to preference for online versus face-to-face instruction. Internet High. Educ. Elsevier **13**, 89–95 (2010)
14. Shen, J., Brdiczka, O., Liu, J.: A study of facebook behavior: what does it tell about your Neuroticism and Extraversion? Comput. Hum. Behav. Elsevier **45**, 32–38 (2015)
15. Bhardwaj, A., Tiwari, A., Krishna, R., Varma, V.: A novel genetic programming approach for epileptic seizure detection. Comput. Methods Programs Biomed. **124**, 2–18 (2016)
16. Bhardwaj, A., Sakalle, A., Chouhan, H., Bhardwaj, H.: Controlling the problem of bloating using stepwise crossover and double mutation technique. Adv. Comput. **2**(6), 59 (2011)
17. Bhardwaj, A., Tiwari, A., Chandarana, D., Babel, D.: A genetically optimized neural network for classification of breast cancer disease. In: 2014 7th International Conference on Biomedical Engineering and Informatics, pp. 693–698. IEEE (2014a)
18. Bhardwaj, H., Sakalle, A., Bhardwaj, A., Tiwari, A.: Classification of electroencephalogram signal for the detection of epilepsy using innovative genetic programming. Expert Syst. **36**(1), e12338 (2019)
19. Bhardwaj, H., Sakalle, A., Bhardwaj, A., Tiwari, A., Verma, M.: Breast cancer diagnosis using simultaneous feature selection and classification: a genetic programming approach. In: 2018 IEEE Symposium Series on Computational Intelligence (SSCI), pp. 2186–2192. IEEE (2018)

20. Bhardwaj, A., Tiwari, A., Varma, M.V., Krishna, M.R.: Classification of EEG signals using a novel genetic programming approach. In: Proceedings of the Companion Publication of the 2014 Annual Conference on Genetic and Evolutionary Computation, pp. 1297–1304. ACM (2014b)
21. Acharya, D., Goel, S., Bhardwaj, H., Sakalle, A., Bhardwaj, A.: A long short term memory deep learning network for the classification of negative emotions using EEG signals. In: 2020 International Joint Conference on Neural Networks (IJCNN), pp. 1–8. IEEE (2020)
22. Bhardwaj, A., Tiwari, A., Vishaal Varma, M., Ramesh Krishna, M.: An analysis of integration of hill climbing in crossover and mutation operation for EEG signal classification. In: Proceedings of the 2015 Annual Conference on Genetic and Evolutionary Computation, pp. 209–216. ACM (2015)
23. Tiwari, S., Goel, S., Bhardwaj, A.: Machine learning approach for the classification of EEG signals of multiple imagery tasks. In: 2020 11th International Conference on Computing, Communication and Networking Technologies (ICCCNT). IEEE (2020)
24. Dodds, K.: Popular geopolitics and audience dispositions: James Bond and the internet movie database (IMDb). Trans. Inst. Br. Geogr. 31(2), 116–130 (2006)
25. Acharya, D., Billimoria, A., Srivastava, N., Goel, S., Bhardwaj, A.: Emotion recognition using Fourier transform and genetic programming. Appl. Acoust. 164, 107260 (2020a)
26. Edla, D.R., Mangalorekar, K., Dhavalikar, G., Dodia, S.: Classification of EEG data for human mental state analysis using Random Forest Classifier. Proc. Comput. Sci. 132, 1523–1532 (2018)
27. Bhardwaj, A., et al.: An innovative genetic programming framework in modelling a real time epileptic seizure detection system (2015)
28. Purohit, A., Bhardwaj, A., Tiwari,A., Choudhari, N.S.: Removing code bloating in crossover operation in genetic programming. In: 2011 International Conference on Recent Trends in Information Technology (ICRTIT), pp. 1126–1130 IEEE (2011)
29. Bhardwaj, A., Tiwari, A., Bhardwaj, H., Bhardwaj, A.: A genetically optimized neural network model for multi-class classification. Expert Syst. Appl. 60, 211–221 (2016)
30. Bhardwaj, A., Tiwari, A.: Performance improvement in genetic programming using modified crossover and node mutation. In: Proceedings of the 15th Annual Conference Companion on Genetic and Evolutionary Computation, pp. 1721–1722 (2013)
31. Bhardwaj, H., Pankaj, D.: A novel genetic programming approach to control bloat using crossover and mutation with intelligence technique. In: 2015 International Conference on Computer, Communication and Control (IC4). IEEE (2015)
32. Acharya, D., Goel, S., Asthana, R., Bhardwaj, A.: A novel fitness function in genetic programming to handle unbalanced emotion recognition data. Pattern Recogn. Lett. 133, 272–279 (2020b
33. Devarriya, D., Gulati, C., Mansharamani, V., Sakalle, A., Bhardwaj, A.: Unbalanced breast cancer data classification using novel fitness functions in genetic programming. Expert Syst. Appl. 140, 112866 (2020)
34. Kumar, A., Sinha, N., Bhardwaj, A.: A novel fitness function in genetic programming for medical data classification. J. Biomed. Inf. 112, 103623 (2020)
35. Acharya, D., et al.: An enhanced fitness function to recognize unbalanced human emotions data. Expert Syst. Appl. 166, 114011
36. Bhardwaj, A., Tiwari, A.: A novel genetic programming based classifier design using a new constructive crossover operator with a local search technique. In: Huang, D.-S., Bevilacqua, V., Figueroa, J.C., Premaratne, P. (eds.) ICIC 2013. LNCS, vol. 7995, pp. 86–95. Springer, Heidelberg (2013). https://doi.org/10.1007/978-3-642-39479-9_11

Smart Specialization Strategies for Smart Cities

The Productivity Forecasting in the Sector of Non-market Services (Education and Health Care Sectors): A Case Study of Ukraine

Svitlana Kozhemiakina[1] , Liudmyla Ilich[1] , Olha Ilyash[2,3(✉)] ,
Svitlana Hrynkevych[4] , and Svitlana Kruvysha[5]

[1] Borys Grinchenko Kyiv University, Kyiv, Ukraine
{s.kozhemiakina,l.ilich}@kubg.edu.ua
[2] International University of Finance, National Technical University of
Ukraine "Igor Sikorsky Kyiv Polytechnic Institute", Kyiv, Ukraine
olha.ilyash@byd.pl
[3] WSG University of Economy in Bydgoszcz, Bydgoszcz, Poland
[4] Lviv Polytechnic National University, Kyiv, Ukraine
svitlana.s.hrynkevych@lpnu.ua
[5] Kyiv Regional Employment Center, Kyiv, Ukraine
svekrivusha@meta.ua

Abstract. Strategic documents lack forecast indicators of the productivity of such non-market sectors of the economy as education and health care. In the article the scientific hypotheses have been generalized. The correlation analysis of forecasting the development of basic non-market economic activities have been proposed. The equations of factor effects on labour productivity in two types of economic activity (education and health care) are developed. A comprehensive econometric model of labour productivity in the sector of non-market services has been suggested. Three blocks of forecast models for estimating the dependence of labour productivity on different factors have been used for calculating the level of growth of labour productivity in the sector of non-market services. The multifactorial dependences of the change in labour productivity are systematized. The obtained quantitative and analytical tools for forecasting labour productivity will improve the effectiveness of state economic policy in non-market services.

Keywords: Productivity forecasting · Labour productivity · Education and health care sectors · Non-market services

1 Introduction

An important component of macroeconomic forecasting is the development of models for forecasting trends in labour productivity growth in Ukraine as a whole, including in the sector of non-market services and its economic activities in order to adopt sound, resource-balanced economic policies. Forecasting the growth of the education and health care economy provides state-level authorities and local governments with

© Springer Nature Switzerland AG 2021
A. Solanki et al. (Eds.): AIS2C2 2021, CCIS 1434, pp. 43–59, 2021.
https://doi.org/10.1007/978-3-030-82322-1_4

a comprehensive tool that combines analysis of current situations, forecasts of socio-economic development and scenarios of the possible future with devising a strategy to achieve it.

Unfortunately, the issues of the labour productivity growth in the sector of non-market services, in particular, in education and health care, have not received due attention, as evidenced by the absence of this indicator in the system of regulation and forecasting of key macroeconomic indicators (KMEI) as well as in the development of sectoral forecasts as required by international methodology. This limits the possibilities of state regulatory influence on the economy's efficiency, the rate and the proportion of economic growth in Ukraine. In this case, the main aim of our paper is to develop methodological approaches to forecasting labour productivity indicators through improving the technology of modelling factor effects on labour productivity and substantiating relationships between indicators-factors taking into account scenario conditions of economic development and the measures of state influence on its level and dynamics in the forecast period.

The study of labour productivity in the sector of non-market services of the economy is carried out in several stages. At the first stage, international scientific achievements in productivity forecasting in general and in the sector of non-market services in particular are analysed and summarised. At the second stage, a factor analysis of labour productivity in non-market services is conducted, and correlations between selected factors and labour productivity are investigated. At the third stage, multifactor models of productivity are built for each of the three blocks of the factors and the statistical significance of these models is checked. At the fourth stage, the forecast of labour productivity in the sector of non-market of the economy is made.

2 Literature Review

The vast majority of research related to productivity has focused on studying labour productivity in material production for a long time. In contrast, the study of factor analysis of productivity of the intangible sphere and its mathematical modelling has not been given enough attention. With the expansion of services and their globalization, scientific research in this area has intensified significantly.

M. Duarte and D. Restuccia investigated the interrelations of sectoral labour productivity in the context of structural economic transformations - secular redistribution of labour by sectors - and the rate of the spread of aggregate productivity in countries. A model of structural transformation was used to determine the sectoral productivity of labour in different countries. The researchers found that the differences in productivity indicators have had an ongoing downward trend in agriculture and industry over the years, while there have been slight changes in the sector of services [9].

In the post-industrial economy, the service industry is developing more dynamically and it is the dominant sector of the economy. The service sector, which is more mobile and sensitive to market needs, can expand the supply of jobs quickly. At present, 60–70% of the employed population of European countries are concentrated in intangible production and the same is true for 75% of the employed population of Canada and the United States. Globalization in the sector of services convincingly proves that for a long time the contribution of this sector to economic growth has been underestimated. In this context, it is worth mentioning the experience of India and China as well. Even though the Chinese economy has long had higher growth rates and prosperity levels than India, India's services have developed at a faster rate. That is why India, by its unique example, has proved to the world that a rapid economic recovery is possible with the expansion of services. However, higher productivity indicators in services compared to industry were another important feature of the Indian economy.

A study by X. Dai, Z. Sun partially confirms this thesis. The study examines the impact of innovations that are being implemented by Chinese manufacturing companies on overall productivity in the industry [5]. This research shows that companies become more productive and attract more resources through innovation, but market distortions greatly influence the redistribution of resources among companies. The research result by X. Dai and Z. Sun suggest that the innovations of Chinese firms did not become a dominant factor in determining the allocation of resources and increasing the overall productivity of the national economy [5]. The study by J. Fagerberg reveals the relationship between the economic structure and labour productivity growth. The scientist studied the dynamics of productivity in 1973–1990, using the data from 39 countries and 24 industries to analyse. The analysis results confirmed the impact of specialization and structural changes on productivity growth in more high-tech industries [11].

Having analysed 12 subsectors of the service industry of 101 countries from 1971 to 2012, B. Kinfemichael and M. Morshed proved that countries, which start to develop in conditions of low initial productivity in the sector of services, are growing much faster than countries with higher initial labour productivity in this sector [18]. I. Kryuchkova suggested considering productivity in the methodological tools of short-term forecasting of key macroeconomic indicators, including changes in the price competitiveness of the country's economy [22]. V. Besedin (2009) proposed the use of a set of multimodel and structural methods, scenario and variant approaches, the method of taking into account external and internal preconditions for forecast periods to predict the industry's productivity through GDP indicator [2].

Sufficient attention is paid to labour productivity problems in economics, which is explained by its stable causal links for any economic system: the higher labour productivity, the greater the country's economic potential is and the higher the welfare of the population becomes. In addition, Ilyash O., Hrynkevych S. and Ilyich L. point out that an increase in the population's well-being is a strategically important direction of modernizing Ukrainian society and improving the quality of life of the population. In this context [16], Ilyash O. (2015) suggests an effective security strategy of the labour market and employment of the population as an immediate goal of productivity growth and social costs reduction through increasing the qualification level and professional development of the workforce and creating the prerequisites for long-term macroeconomic stabilization [17]. It is also worth mentioning the study by L. Ilich, O. Hlushak and S. Semenyaka, which is devoted to modelling structural changes in the labour market for investigating the dependence of the employment level of the population in the regions on the availability of higher education, growth rates of labour productivity, average wages, a capital investment index, the value of the export-import coverage ratio [15].

In the economic research of the 1990s, one can trace the arguments in favour of the relationship between standards of living and productivity. A. Blinder and J. Yellen proved that productivity growth has a positive effect on economic development [3]. P. Krugman scientifically substantiated the thesis that the growing dynamics of productivity are extremely important in the long run, affecting the country's welfare [21]. S. Haller and S. Lyons suggested that the rapid growth of labour productivity in Asia's service sector brings benefits to all sectors of the economy and contributes to the sustainable and balanced growth of Asian economies. The scientists also noted that the rapid productivity growth in services is accompanied by an expansion of material production, which is necessary for ensuring fixed capital, which will accelerate economic growth [12].

L. Dearden, H. Reed, and J. Reenen established a direct link between on-the-job training and productivity. In particular, for a panel of British industries, the authors find that a 1% increase in work-training rises about 0.6% the value-added per hour and about 0.3% the hourly wage [6]. Hrynkevych S. proved the impact of the educational component and health component on the growth of the efficiency of the labour potential use [14]. N. Azenui and C. Rada found that global integration through trade has a small effect on productivity growth in the industry and structural change. At the same time, the researchers noticed that production and foreign direct investment are linked to a rise in labour productivity and its components at both the sectoral and national levels [1].

After examining the relationship between education, productivity, and wages at the company level in all Italian economy sectors between 1996 and 1999, G. Conti, and wages at the company level in all Italian economy sectors between 1996 and 1999, G. Conti found that education significantly raises productivity. However, with regard to wages, such an effect is not observed. This confirmed the fact that companies appropriate most of the profits [4]. H. Sala and J. Silva noticed a link between professional education and labour productivity growth in Europe [26]. The authors proved that one extra hour of training per employee accelerates productivity growth by 0.55 percentage points.

Despite significant scientific developments in this area, there is almost no research to assess productivity in education and health care. In post-socialist countries, a simplified approach to substantiating the future dynamics of the health care and education sectors and the labour and capital resources prevails. The main directions of reforming these sectors consist in regarding the fields of education and health care exclusively as consumers of resources and not as producers. This approach negatively affects the competitiveness of the labour force and the economic development of post-Soviet countries. As a result, these sectors experience chronic insufficiency of state funding; there are attempts to increase the involvement of people in the financing of health care and education without taking into account their solvency.

3 Methodology

Based on the theory and methodology of studying the interconnections and interdependence of phenomena and processes, it should be added that: 1) Performance indicators of the health care and education sectors are formed under the influence of a set of interrelated factors of a macro- and microeconomic level, which function differently. 2) The difficulty of forecasting and modelling consists of choosing the factors that would be the most informative and reflect objectively socio-economic processes of the impact on labour productivity. In this case, correlation analysis is a more reasonable method for selecting factors, as it allows us to quantify the statistical relationship between indicators, which exists in the reporting period [25].

In general, the algorithm of forming a quantitative and analytical tool for forecasting labour productivity in the sector of non-market services involves the following sequence of steps: 1) construction of an econometric model of increasing labour productivity in the sector of non-market services. The process of econometric modelling includes the identification of a system of labour productivity indicators in the sector of non-market services, the determination of the relationships between factors (based on a matrix of paired correlation coefficients), the formation of approaches (blocks of the model) to assessing the dependence of labour productivity on factors (based on a multifactor equation) and the modelling of structural changes in the sector of non-market services. The first block of the model reflects the impact of macroeconomic parameters on labour productivity in the sector of non-market services; the second block of the model reflects the impact of certain economic activities of the sector of non-market services (education and health); the third one – the impact of time changes on labour productivity (in a generalized form, time changes are identified with the action of the aggregate factor of scientific and technological progress and the development of innovation in industries); 2) development of technology of using the model for studying labour productivity. The first stage involves forecasting the coefficients of the closeness of the dependence of labour productivity in the sector of non-market services on the model's factors. The second stage includes the formation of major trends in labour productivity changes in the sector non-market services depending on the factor impact. The third stage is associated with the study of the impact of regulatory factors on labour productivity. 3) coordination, evaluation, and calculation of the forecast indicators of labour productivity in the sector of non-market services, coordination, evaluation, and expert adjustment.

In the sector of non-market services (Y) as whole labour productivity in health care (Y^1) labour productivity in education (Y^2) the labour productivity is used. Factors (x) that affect labour productivity in this sector include the following macroeconomic indicators: x^1- capital investment, UAH billion; x^2- GDP (at previous year's prices), %; x^3- average monthly wages (nominal); x^4- the number of the employed population, thousand people; x^5 - consumer price index (from December to December),%; x^6- a consolidated budget (revenues), UAH billion; x^7- a consolidated budget (expenditures), UAH billion; x^8- the number of permanent population (at the end of the year), million people; x^9- unemployment rate (according to the ILO methodology),%; x^{10} – the number of the unemployed (according to the ILO methodology), thousand people; x^{11} - GDP (at actual prices), UAH billion; x^{12} - average monthly wages, UAH; x^{13}- the number of teachers of educational institutions, thousand people; x^{14}- total costs per pupil/student, UAH; x^{15} – the number of educational institutions, thousand units; x^{16}- the number of pupils/students, thousand people; x^{17}- average monthly wages, UAH; x^{18} – the number of employed medical staff, thousand people; x^{19}- the number of hospital beds, thousand units; x^{20}- total health care costs per capita, UAH; x^{21}- labour productivity in education at actual prices (UAH/person); x^{22}- labour productivity in health care and social assistance at actual prices (UAH/person); x^{23}- time (the index of the year).

The first block of the model considers factors from x^1 to x^{11} inclusive. Two dependence equations are constructed on this factor base. In addition to the multifactor equation of the mutual influence of microeconomic factors on labour productivity in education and health care, the second block of the model uses labour productivity indicators in education (x^{21}) and health care (x^{22}), which depend on public funding as factors of influence (x) on labour productivity in the sector of non-market services. Model constructions are based on the analysis of the economic situation in the reporting period, first of all, on the study of the dynamics of labour productivity and the factors influencing it. Labour productivity in education and health care (these types of economic activity constitute 60% of non-market services) directly affects labour productivity in the sector of non-market services as a whole. It is a significant factor in changing this indicator. A factor base from x^{12} to x^{16} inclusive was used to construct the equation of dependence of labour productivity in education and a factor base from x^{17} to x^{20} inclusive was applied when constructing the same equation of labour productivity in health care. The third block of the model reflects the dependence of labour productivity on the generalized impact of positive innovative, scientific, and technical changes in providing non-market services through the time factor (x^{23}). The formalized dependence of labour productivity on the time factor is given.

In general, the proposed quantitative and analytical tools for forecasting labour productivity in the sector of non-market services will serve as an effective means of developing and adjusting the state's economic policy.

4   Findings and Discussion

To assess the closeness of statistical dependence, a group of essential factors was selected and included in the multifactorial equation of dependence (second block, formula 3) of the labour productivity indicator on those factors. This group includes nominal wages, UAH; the number of the employed population, thousand people; the number of hospital beds, thousand; total health care costs per capita, UAH. Thus, this dependence indicates that wages, the number of the employed population, the number of hospital beds, total health care costs per capita have the most significant impact on productivity in health care. The result of our calculations shows that a rise in wages by one hryvnia increases the level of labour productivity by 3.49 hryvnias per person. An increase in the number of those who work in health care by 1,000 people raises labour productivity by 62 hryvnias per person. In contrast, a decrease in the number of hospital beds by 1000 units reduces productivity by 54 hryvnias per person. A rise in society's total expenditure on health care per person by 1 hryvnia increases labour productivity by UAH 18 hryvnias per person.

Similarly, to build a «sectoral» model of changes in labour productivity in education, the typical microeconomic factors are being studied: capital investment, UAH million; capital-labour ratio, %; nominal wages, UAH; the number of the employed population, thousand people; the number of teaching staff, thousand people; total costs per pupil/student, UAH; the number of educational institutions, thousand units; the number of pupils/students, thousand people; total expenditures on education, UAH million; total expenditures on education, % of GDP; government spending on education, % of GDP;

gross value added GVA (at actual prices), UAH million; an index of fixed assets growth (%).

Using the matrix of paired correlation coefficients, the most significant of them are determined and included in the multifactorial equation of the mutual influence of factors on labour productivity in education (second block, formula 3): nominal wages, UAH; the number of teaching staff, thousand people; total costs per pupil/student, UAH; the number of pupils/students, thousand people. The conclusion that follows from this dependence is that a rise in wages by one hryvnia increases labour productivity by 13.8 hryvnias. A reduction in the number of teachers by 1,000 people increases labour productivity by 130.9 hryvnias per person, an increase in total costs per pupil/ student by 1 hryvnia reduces labour productivity by 1.57 hryvnias per person, a rise in the number of pupils/students by 1,000 people increases labour productivity by 32.6 hryvnias per person. The equations show that there is a correlation between the dependent variable (labour productivity) and the factors that affect it, and also explain by 97 and 99% the change in the resultant indicator – the coefficient of determination (R2) is 0.97 and 0.99.

The developed equations of factor influences on labour productivity in the main economic activities (education and health care) of the sector of non-market services, unfortunately, do not provide adequate values of labour productivity, probably due to multicollinearity – the presence of a linear relationship between independent variables, which are included into a complex model as additional blocks to strengthen its charac-teristics and improve the quality of calculations. Another disadvantage of this method is the lack of significant macroeconomic regressors.

A formalized description of factor influences on changes in labour productivity in the sector of non-market services is presented in three blocks of models:

The first block: The equation of the influence of selected factors on productivity in the sector of non-market services (Coefficient of determination R = 0.95)

$$1 - a)Y = -1063, 6x_2 - 420, 87x_{5+}83, 79x_6 + 4936, 1x_9, \tag{1}$$

$$1 - \sigma)Y - 98204, 9 - 122, 734x_1 + 3, 4940x_{4+} 29, 93314x_7 + 7, 268416x_{10} + 52, 6973x_{11} \tag{2}$$

The second block: The equation of the influence of selected factors on labour productivity in education (Coefficient 3 of determination R = 0.99):

$$Y^1 = -99426, 71 + 13, 89x_{12} + 130, 85x_{13} - 1, 57x_{14} - 14055, 4x_{15} + 32.60x_{16} \tag{3}$$

The equation of the influence of selected factors on labour productivity in health care (Coefficient of determination R = 0.97):

$$Y^2 = -71715, 88 + 3, 49x_{17} + 62, 42x_{18} - 54, 81x_{19} + 18, 19x_{20} \tag{4}$$

The equation of the influence of selected factors on labour productivity in the sector of non-market services (Coefficient of determination R = 0.99):

$$Y = -14603, 5 + 1, 185394x_{21} + 0, 239x_{22, } \tag{5}$$

The third block: The equation of the influence of factors on labour productivity in the sector of non-market services (Coefficient of determination R = 0.96):

$$Y = 21079, 58 + 7174, 852x_{23, } \tag{6}$$

The proposed multifactor model dependences of labour productivity dynamics on selected factors allow factor relationships to be classified and the most important ones to be determined. It should be noted that all factor indicators that were included in the equations of the model have a different direction vector (increase or decrease) and a different quantitative influence on the value of labour productivity in the sector of non-market services. The coefficients, which are found in the equations at the necessary factor indicators, show the share of the change introduced by this factor to labour productivity for a given value of the factor indicator and this share depends on the indicator factor itself in the year under study. Thus, the contribution of the factor to the value of the Y model (labour productivity in the sector of non-market services) is the product of the value of the factor indicator and its coefficient. To compare the impact of the factors on labour productivity among themselves, it is advisable to take one percent of this value, i.e. to estimate the increase (decrease) of the target indicator Y (labour productivity in thousands of UAH) when changing each individual factor (x_1) by one %. The results of such calculations are shown in Table 1.

Table 1. The rating of the factors by their influence on labour productivity in the sector of non-market services

Indicator of the factor	Change in labour productivity when the factor changes by one %	Rating (the place by the impact on labour productivity) of the factor
Number of teachers in educational institutions, thousand people	39,1	7
Number of people employed in health care, thousand people	48,6	5
Number of hospital beds, thousand	12,2	9
Number of pupils/students, thousand people	85,2	3
Total health care costs, thousand UAH per person	62,5	4
Wages in education (nominal), UAH	44,3	6
Time (innovative changes)	110,4	2
Number of the unemployed (according to the ILO methodology), thousand people	121,8	1
The average monthly wages (nominal) in health care, UAH	12,0	10

(continued)

Table 1. (*continued*)

Indicator of the factor	Change in labour productivity when the factor changes by one %	Rating (the place by the impact on labour productivity) of the factor
Total costs per pupil/ student, UAH	17,0	8
Unemployment rate (according to the ILO methodology), percent	0,11	16
GDP growth rate (at previous year's prices),%	0,80	14
Consumer price index (from December to December),%	0.34	16
Consolidated budget revenues, UAH billion	1,26	11
Capital investment, UAH billion	0,20	16
GDP (at actual prices), UAH billion	1,23	12
Labour productivity in education	**0,90**	**13**
Labour productivity in health care and social assistance	**0,67**	**15**
Number of the employed population, thousand people	0,30	16
Consolidated budget expenditures, UAH billion	0,25	16

The system of the factors by the power of their influence on labour productivity is given in the form of a special rating depending on the data in column 2 of Table 1. The power of the factor influence on labour productivity was assessed regardless of the direct impact (increase or decrease). When forecasting the labour productivity indicator, one should take into account the leading multidirectional socio-economic trends, which are equally crucial for ensuring the growth of labour efficiency. For instance, a constant reduction in the number of the population as a whole is a stable demographic process that will lead to a decrease in employment in Ukraine and the sector of non-market services in particular. Along with the population decline, the number of non-market services recipients, including those from education and health care, will decrease. It reduces the result of their activity in the form of value-added, which is created in the sector of non-market services and these things also affect the denominator when calculating the level of labour productivity.

According to the model equations' assessment, the number of the unemployed, the level of innovation and technology, the number of pupils/students and total financing costs to be the most significant factors influencing the level of labour productivity (y) in health care. Several factors included in the model have a small impact and, accordingly, a relatively low rating. However, their total single-vector effect is sufficient for them to be reflected in the model for predicting labour productivity. The current scheme of the labour productivity model in the sector of non-market services shows an assessment of productivity dynamics when changing the relevant factors. The forecast has several stages, including the stage of a preliminary forecast of the factors included in the model equation. The corresponding time series of indicators-factors are shown in Table 2.

Table 2. Reporting and forecasting data of macroeconomic factor indicators for the first block of the model of predicting labour productivity in the sector of non-market services in 2020–2023.

Indicators	2010	2015	2017	2018	2019 Expec.	2020 Pred.	2021 Pred.	2022 Pred.	2023 Pred.
Gross domestic product (at actual prices), UAH billion	1120,6	1988,5	2983,2	3558,7	3610	**3960**	**4325**	**4740**	**5200**
GDP index (at previous year's prices)	104,1	90,2	102,5	103,3	103,0	**103,8**	**104,1**	**104,3**	**104,5**
Consumer price index (from December to December of the previous year), percent	109,1	143,3	113,7	109,8	107,4	**105,6**	**105**	**105**	**105**
Capital investment, UAH billion	181,1	273,1	448,5	578,7	542	**602**	**666**	**749**	**842**
Number of the employed population, thousand people	20266	16443,2	16156	16361	16117	**16102**	**16092**	**16087**	**16087**
Consolidated budget (revenues), UAH billion	314,5	652	1017	1184	1137	**1236**	**1341**	**1470**	**1612**
Consolidated budget (expenditures), UAH billion	377,8	679,9	1057	1250	1264	**1354**	**1471**	**1612**	**1768**

(*continued*)

Table 2. (*continued*)

Indicators	2010	2015	2017	2018	2019 Expec.	2020 Pred.	2021 Pred.	2022 Pred.	2023 Pred.
Unemployment rate (according to the ILO methodology), percent	8,1	9,1	9,5	8,8	9,5	**9,4**	**9,3**	**9,2**	**9,0**
Number of the unemployed (according to ILO methodology), thousand	1785,6	1654,7	1698,0	1578,6	1692	**1671**	**1650**	**1630**	**1591**

For the first block of the model of forecasting macroeconomic factors, the official forecast on socio-economic development of Ukraine, prepared by the Cabinet of Ministers for the budget for 2019 and approved by resolutions of the Cabinet of Ministers dated 01.01.2017 № 411 «On approval of the forecast of economic and social development of Ukraine for 2018–2020» and by resolution of the Cabinet of Ministers of Ukraine № 546 dated 11.07. 2018 «On approval of the forecast of economic and social development of Ukraine for 2019–2021» [26] were used.

The forecast takes into account the trends and assumptions formed for the post-crisis period from 2016 to 2018 with a slight improvement in the medium term: GDP growth (in comparable prices) at a rate of 103–104.5% per year; a significant reduction of inflation to 5% in 2023; the growth of the share of capital investments in GDP from 15% in 2019 to 16.1% in 2023; a slow reduction of unemployment by 0.5 percentage points for 5 years; a decrease in the number of the employed due to expected demographic factors; the relatively optimal ratio between GDP and budget revenues (31–32%).

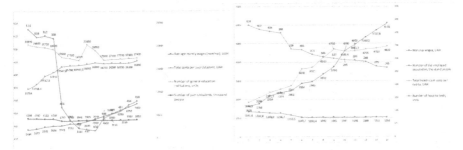

Fig. 1. Reporting and forecasting data of factor indicators for the second block of the model of forecasting labour productivity in education in 2020–2023.

Fig. 2. Reporting and forecasting data of factor indicators for the second block of the model of forecasting labour productivity in health care in 2020–2023.

To calculate labour productivity in the first block of the model (the impact of macroeconomic factors) the employed the macroeconomic forecast shown in Figs. 1 and 2 (1-a and 1-b) is employed. As a result of the calculations based on the equations of the model for education and health care (formulas 1–5), the forecast of factor indicators for the second block of the model has been developed, which is given in Fig. 3.

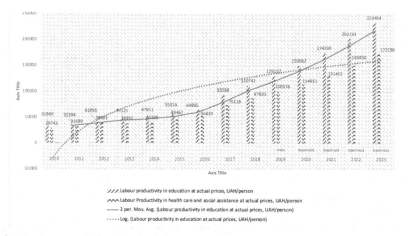

Fig. 3. The forecast of labour productivity in education and health care as factor indicators for the second block of the model of predicting labour productivity in the sector of non-market services in 2020–2023.

Table 3. The forecast of labour productivity in the sector of non-market services by three scenarios in 2019–2023 (thousand UAH/person)

	Labour productivity for the reporting period	Labour productivity in the sector by the 1st scenario of the model (1-a and 1-b)		Coefficient of deviation		Labour productivity by the 2nd scenario of the model,	Coefficient of deviation	Labour productivity by the 3rd scenario of the model	Coefficient of deviation
2010	30,9	27,2	30,1	0,12	0,03	31,0	0,00	19,1	0,38
2011	35,4	36,9	38,5	−0,04	−0,09	35,7	−0,01	30,5	0,14
2012	43	47,0	43,4	−0,10	−0,01	45,4	−0,06	41,9	0,02
2013	47,4	46,3	47,0	0,02	0,01	48,0	−0,01	53,4	−0,13
2014	52,8	47,6	47,6	0,10	0,10	50,5	0,04	64,8	−0,23
2015	63	65,8	63,5	−0,04	−0,01	62,5	0,01	76,2	−0,21
2016	76,2	80,9	80,2	−0,06	−0,05	73,7	0,03	87,6	−0,15
2017	104,3	106,5	104,0	-0,02	0,003	107,6	-0,03	99,0	0,05
2018	129,9	124,8	128,6	0,04	0,01	128,6	0,01	110,4	0,15
2019	-	121,4	128,9	-	-	150,7	-	121,8	-
2020	-	132,2	143,1	-	-	175,6	-	133,2	-
2021	-	143,7	157,9	-	-	204,4	-	144,6	-
2022	-	157,7	175,2	-	-	237,5	-	156,1	-
2023	-	173,0	194,4	-	-	275,0	-	167,5	-

The data in Fig. 4 indicate that the projected level of labour productivity in education and health care is used as a factor indicator for calculating labour productivity in the sector of non-market services (the second block of the model). The forecast of labour productivity in non-market services by three blocks of the model and the average aggregate forecast are presented in Tables 3 and 4 and Fig. 4.

Based on the analysis of the deviations of the actual trend of labour productivity for the reporting period (2010–2018) and the calculated trends in different scenarios, it is found that the most severe errors occur in block 3. From the data of Tables 3 and Fig. 4, one can see that different blocks of the model in different ways reflect changes in the level of labour productivity in the medium term. A more optimistic forecast was obtained for the second block of the model (the equation of the first block of the model) (a macroeconomic impact). Then, it is proposed to consider block 1-a, block 1-b and block 2. The smallest deviations are observed in block 2 (based on the impact of labour productivity in education and health care).

Table 4. The generalized forecast of labour productivity in the sector of non-market services (thousand UAH /person)

	The specific weight of the first scenario in the overall forecast, %		The contribution to the generalized forecast made by the forecast according to the first scenario		The specific weight of the second scenario in the overall forecast, %	The contribution to the generalized forecast made by the forecast according to the second scenario	The specific weight of the third scenario, %	The contribution to the generalized forecast made by the forecast according to the third scenario	The generalized forecast by the three Scenarios
	Scenario 1-a dla = 17,3%	Scenario 1-b d1b = 29,9%	Scenario 1-a	Scenario 1-b	Scenario 2 d2 = 46,3%		Scenario 3 d3 = 6,4%		Weighted average assessment by 3 scenarios
2010	-	-	-	-	-	-	-	-	30,9
2011	-	-	-	-	-	-	-	-	35,4
2012	-	-	-	-	-	-	-	-	43,0
2013	-	-	-	-	-	-	-	-	47,4
2014	-	-	-	-	-	-	-	-	52,8
2015	-	-	-	-	-	-	-	-	63,0
2016	-	-	-	-	-	-	-	-	76,2
2017	-	-	-	-	-	-	-	-	104,3
2018	-	-			-		-		129,9
2019 Expec	-	-	21,0	38,5	-	69,8	-	7,80	137,11
2020	-	-	22,9	42,8	-	81,3	-	8,52	155,49
2021	-	-	24,9	47,2	-	94,6	-	9,25	175,96
2022	-	-	27,3	52,4	-	110,0	-	9,99	199,62
2023	-	-	29,9	58,1	-	127,3	-	10,72	226,10

*2020–2023 – prediction periods

Accordingly, if the errors are combined into the solo forecast of the non-market services sector labour productivity, the deviations have an inversely proportional share in the all blocks' errors terms (according to the blocks, these are weighting factors - 0.252, 0.222, 0.330 and 0.175).

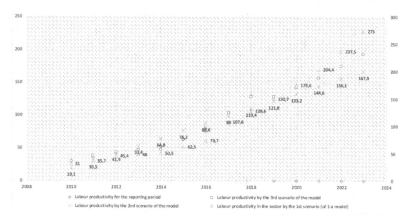

Fig. 4. The forecast of labour productivity until 2023 in the sector of non-market services by the three scenarios of the model (thousand, UAH)

The general forecast shown in Fig. 5 is constructed as a weighted average of the forecast calculations for different model blocks. The forecast for the third block is more pessimistic (the forecast by time factor, or taking into account scientific, technical and innovative trends in the sector of non-market services), which is explained by a generally low level of innovation of the Ukrainian economy and a low impact of this factor in the sector of non-market services.

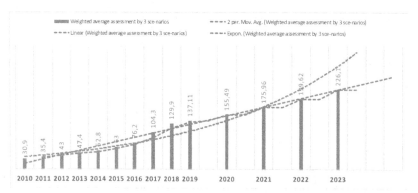

Fig. 5. The generalized weighted average forecast of labour productivity until 2023 in the sector of non-market services of the economy of Ukraine by the three blocks of the model (thousand, UAH)

The first equation of the first block and of the second block of the model demonstrates the most average and relative values of the forecast level of labour productivity. The general average forecast of labour productivity in the studied sector is close to labour productivity in the second block of the model.

The equations of the model allow us to predict the level of labour productivity at current prices. To recalculate these indicators compared to the price, a deflator in the form of a consumer price index, which is projected in Table 2 is used. Under the scenario conditions and assumptions built in the developed forecast, the real growth rate of labour productivity will rise annually from 2.0% in 2019 up to 3.5% in 2023.

5 Conclusions

Within the framework of our research, the model of forming quantitative and analytical tools for forecasting labour productivity in the sector of non-market services through improving the technology of modelling factor influences on labour productivity, substantiating interrelations between the indicators-factors together with taking into account the scenario conditions of economic development as well as measures of state influence on its level and dynamics in the forecast period have been proposed and substantiated.

The model includes an algorithm for building an econometric model of increasing the level of labour productivity in the sector of non-market services, the technology of using the model for studying labour productivity, and the calculation of the forecast of labour productivity in the sector of non-market services.

Considering the existing trends and the factors used in the model, the forecast calculations have shown that the growth rate of labour productivity of the sector of non-market services will significantly lag behind the growth rate of labour productivity of the economy as a whole. To equalise these rates, it is necessary to provide an increase in the share of public funding in the development of education and health care and achieve a higher level of efficiency of budget financing.

The overall level of commercialization of the sector of non-market services will go up slowly and its impact on labour productivity in the medium term will be insignificant. To implement the developed forecasts of labour productivity in education and health care based on the proposed methodological approaches, further research should focus on the main directions of government influence on labour productivity growth in the sector of non-market services. Those directions must aim to improve the results of general character functioning and increase the efficiency of allocating labour and capital resources.

In this article, the quantitative and analytical tools for forecasting labour productivity in the national economy for the prediction of state economic policy effectiveness of public authorities by assessing the future dynamics of labour productivity in the sector of non-market services in the structure of key macroeconomic indicators of the national economy are suggested. However, in future research, the dynamics of labour productivity in the sector of non-market services for the medium-term forecast period, which can be calculated based on the multifactor regression equation in different scenarios (optimistic, pessimistic, crisis, etc.) are explored. With the development of appropriate external and internal assumptions and risks. Moreover, it reasonable to compare the calculations' results in terms of intensity and deviations in growth rates.

It is that in future research, to improve the forecast developments in education and health care based on the proposed methodological approaches, it is advisable to focus on substantiating the main directions of state regulation of processes and conditions of labour productivity growth in the sector. In our opinion, these directions should improve the effectiveness of the sector of non-market services and the efficiency of allocating labour and capital resources as leading factors in the market economy growth.

References

1. Azenui, N.B., Rada, C.: Labor productivity growth in sub-Sahara African LDCs: sectoral contributions and macroeconomic factors. Struct. Change Econ. Dynamis **56**, 10–26 (2021). https://doi.org/10.1016/j.strueco.2020.07.005
2. Besedin, B.: Gross industry product as an indicator of the volume of the inland product of the industry and its productivity. Ukraine Econ. **1**, 31–42 (2009)
3. Blinder, A., Yellen, J.: The Fabulous Decade: Macroeconomic Lessons from the 1990s (2001)
4. Conti, G.: Training, productivity, and wages in Italy. Labour Econ. **12**, 557–576 (2005). https://doi.org/10.1016/j.labeco.2005.05.007
5. Dai, X., Sun, Z.: Does firm innovation improve aggregate industry productivity? Evidence from Chinese manufacturing firms. Struct. Change Econ. Dyn. **56**, 1–9 (2021). https://doi.org/10.1016/j.strueco.2020.09.005
6. Dearden, L., Reenen, J., Reed, H.: The impact of training on productivity and wages: evidence from British panel data. Oxford Bull. Econ. Stat. **68**, 397–421 (2006). https://doi.org/10.1111/j.1468-0084.2006.00170.x
7. Donchenko, V., Tarasova, O.: Matrix multiple regression. Bulletin of the Taras Shevchenko National University. Ser.: Phys. -Math. Sci. **2**, 133–138 (2015)
8. Donchenko, V., Zinko, T., Skotarenko, V.: The concept of a tuple for linear operators and its implementation for matrix tuples. J. Comput. Appl. Math. **3**(120), 127–140 (2015)
9. Duarte, M., Restuccia, D.: The role of the structural transformation in aggregate productivity. Q. J. Econ. **125**, 129–173 (2010). https://doi.org/10.1162/qjec.2010.125.1.129
10. Emelianenko, L., Radionova, I., Fedirko, N.: Macroeconomic assessment of the role of the public sector of the national economy and its deviations in Ukraine from general civilization trends: analytical note on the results of research work. Crisis management of the public sector in macroeconomic instability and threats to the statehood of Ukraine (state registration number: 0117 U 001196). KNEU, Kyiv (2017)
11. Fagerberg, J.: Technological progress, structural change, and productivity growth: a comparative study. Struct. Change Econ. Dyn. **11**(4), 393–411 (2000). https://doi.org/10.1016/S0954-349X(00)00025-4
12. Haller, S.A., Lyons, S.: Effects of broadband availability on total factor productivity in service sector firms: evidence from Ireland. Telecommun. Policy **43**, 11–22 (2019). https://doi.org/10.1016/j.telpol.2018.09.005
13. Heitz, V. (ed.): Economy of Ukraine: Strategy and Policy of Long-Term Development. Kyiv, Phoenix (2003)
14. Hrynkevych, S., Gural, N.: Assessment of labour potential renovation for trade enterprises. Econ. Annals-XXI **160**(7–8), 96–99 (2016)
15. Ilich, L. Hlushak, O.M., Semenyaka, S.A.: Modeling of employment structural transformations. Financ. Credit Act. Problems Theor. Pract. **1**(32), 251–259 (2020). https://doi.org/10.18371/fcaptp.v1i32.200469

16. Ilyash, O., Hrynkevych, S., Ilich, L., Kozlovskyi, S., Buhaichuk, N.: Economic assessment of the relationship between housing and communal infrastructure development factors and population quality of life in Ukraine. Monten. J. Econ. **16**(3), 93–108 (2020). https://doi.org/10.14254/1800-5845/2020.16-3.8

17. Ilyash, O.: Strategic priorities of Ukraine's social security concept development and implementation. Econ. Ann. XXI **7–8**, 20–23 (2015)

18. Kinfemichael, B., Morshed, A.K.M.M.: Unconditional convergence of labor productivity in the service sector. J. Macroecon. **59**, 217–229 (2019). https://doi.org/10.1016/j.jmacro.2018.12.005

19. Kocharyan, I., Goritsyna, I., Klymenyuk, M.: Forecasting the use of higher education potential in the post-crisis economy of Ukraine. Bulletin of the National .University of Water Management and Environmental Sciences: Coll. Science. Wash. Ser. Econ. **4**(3), 268–273 (2004)

20. Kolbasynsky, S.: Analysis of the economic and mathematical apparatus for modeling and forecasting of state budget performance indicators and macroeconomic indicators. Economic Bulletin of the National Technical University of Ukraine, Kyiv Polytechnic Institute, 12, (2015). https://doi.org/10.20535/2307-5651.12.2015.44169

21. Krugman, P.: The Age of Diminished Expectations: U.S. Economic Policy in the 1990s. MIT Press, Cambridge, Mass (1994)

22. Kryuchkova, I. (ed.): Macroeconomic Modeling and Short-Term Forecasting. Kyiv (2004)

23. Levytska O.O., Mulska O.P., Ivanyuk U., Vasyltsiv T.G., Kunytska-Iliash M., Lupak R.L.: Modelling the conditions affecting population migration activity in the Eastern European region: the case of Ukraine. TEM J. **9**(2), 507–514. (2020) (Scopus). https://doi.org/10.18421/TEM92-12

24. Mulska, O., Levytska, O., Panchenko, V., Kohut, M., Vasyltsiv, T: Causality of external population migration intensity and regional socio-economic development of Ukraine. Probl. Perspect. Manag. **18**(3), 426–437. (2020) (Scopus) https://doi.org/10.21511/ppm.18(3).2020.35

25. On approval of the Forecast of economic and social development of Ukraine for 2019–2021: Resolution of the Cabinet of Ministers of Ukraine dated 11 July 2018 № 546, https://zakon.rada.gov.ua/go/546-2018-%D0%BF. Accessed 17 Nov /2020

26. Sala, H., Silva, J.: Labor productivity and vocational training: evidence from Europe. IZA Discussion Paper **6171**, 1–22 (2013)

An Emergent Role of Knowledge Graph and Summarization Methodology to Simplify Recruitment for the Indian IT Industry

Anshul Ujlayan[1]([⊠]) [iD] and Manisha Sharma[2] [iD]

[1] School of Management, Gautam Buddha University, Greater Noida, India
[2] School of Business Management, NMIMS, Mumbai, India
manisha.sharma@sbm.nmims.edu

Abstract. In the new age of information revolution, recruiters are getting many job applicants' profiles from various sources. Recruiters invest considerable time and effort to evaluate and organize this amount of data in semi-structured or unstructured format in the information technologies industry. To understand and summarize job applicants' profiles, a knowledge graph can help to provide instant screening. This paper proposes the use of a machine learning techniques for the generation of knowledge graph and extractive summarization of job applicants. This will help to have a quick knowledge graph visualization and short summary of relevant information from candidates' profiles. The results of the study can significantly reduce the effort and time taken to manually screen profiles for matching jobs during a recruitment process.

Keywords: Knowledge graph · Machine learning · Recruitment process · Extractive summarization

1 Introduction

In the IT industry, all companies and recruitment agencies process and screen numerous resumes daily. Therefore, with the rise in digital information, as quick, and efficient automated summarization, it has become more important than ever before. The recruitment team in various organizations, are working to invest minimum time to get maximum output. The purpose is to get a detailed view of applicants' profiles befitting the job description. Most of the recruiters expend a great deal of time, effort and money during the screening process to come up with a shortlist of applicants with a profile summary. There is a need for an automated intelligent analytics process to minimize time and effort, pulling out all the critical data from the unstructured resumes to prepare the summary.

There are various ways to summarize candidates' profiles. Sentence extraction is used by most existing automated summary systems where the main sentences in the given document are extracted to create a summary. Most of the scoring techniques for text summarization use both strictly statistical and exclusively semantic characteristics [17].

A. Solanki et al. (Eds.): AIS2C2 2021, CCIS 1434, pp. 60–72, 2021.
https://doi.org/10.1007/978-3-030-82322-1_5

Due to a large amount of available data, it is difficult for individuals to interact with the necessary information. Thus, new technologies and techniques based on Natural Language Processing are crucial for the efficient processing of all information. In this respect, Information Extraction and Text Summarization will make it easier for recruiters to read the information, mostly by minimizing the time they consume dealing with the data and choosing the most valuable details to them [1]. It is tedious to sort out many applicants' profiles from a huge set of applicant's profiles [6]. It can take minutes without a review or rundown, to make sense of what people would address in a document or article. The automated text summarizer that concentrates a sentence from a record of material offers the most essential information and returns it in a readable and ordered manner. Some of the aspects in which the use of artificial intelligence, like automated recruiting practices, peer mentoring processes, information performance assessments and predictable productivity and attrition, can be improved and streamlined. Human Resource (HR) Managers can use artificial intelligence features such as machine learning, and natural language processing [21].

Artificial intelligence offers great opportunities to elevate and automate HR functionality to help organizations to achieve their goals in less time. With the increase in technology consumption, organizations will be requiring highly skilled professionals who can make the machines perform tasks as per the requirement. AI will help employees to manage their work-life balance effectively. They will complete their tasks before deadlines, thereby reducing the dependency on employees in the organization [16].

2 Objectives of the Study

Based on the literature review, the research focus is on the recruitment process for the Indian IT industry. The study proceeds to lay down the objectives as below:

1. Candidates' profile summarization to help the recruitment analyst with a quick overview using the extractive summarization approach.
2. Knowledge Graph visualization for candidates' profiles to reduce the effort and time to look at the profile characteristics.

3 Literature Review

A detailed literature review is conducted for the current study that suggests the use of technology like natural language process and machine learning. These advance technologies can help recruiters to perform candidates' profile screening rapidly [4]. There are different classification methods that are tailored to the candidates' profile summary task and evaluate their success based on human validation trials and a real-life data collection of news articles [11].

The method of processing vast volumes of text to derive high-quality information is text mining. To carry out text analysis, it implements some of the natural language processing techniques such as parts-of-speech marking, parsing, N-grams, tokenization, etc. Tasks such as automated extraction of keywords and text summarization are included. It is possible to achieve the text summary in two ways: abstractive summary and extractive summary. The abstract description is a matter of immense scope for study. An extractive description removes material from the main report itself and introduces it to the reader [3].

The task is to find the few terms and phrases that best characterize or differentiate the subject as it appears in the documents, provided a corpus of documents, and a subject of interest i.e., represented as a shortlist of words. To reduce several millions of words into a few indicative main phrases use scalable, reproducible machine learning techniques. Consider these phrase lists as summaries of how the subject in the corpus is handled [5]. By topic, in a collection of text documents such as country, person, economy, etc., we mean a noun, topic, or theme of interest, and by treatment, it means how a collection of documents addresses a particular subject and in what context. In this case, a summarizer is an automated method that takes a series of documents and a topic of interest and returns the description, i.e., a list of key phrases explaining how the topic is addressed throughout the documents.

Choosing the top representative sentences from the input document is the key challenge of extractive-base text summarization. Several approaches, such as feature-base, cluster-base, and graph-based methods, have been proposed to improve the selection process. This paper basically proposes to improve previous work, include some limitations on summarization approaches, and provide improved mixed feature-based and cluster-base approaches to generate a highly qualified single-document overview summary [1].

For almost the last half-century, the natural language processing group has studied the sub-field of summarization. A summary was defined by [18] as a text created from one or more texts, conveying significant information in the original text(s), which is no more than half of the original text and typically significantly less than that. Three significant aspects that characterize automatic summary research are captured by this simple description and form the basis for this work.

The knowledge graphs represent significant entities' common representations and relationships. Graphs can promote human data analysis as well as enable memory-dependent information-based applications through lightweight, interpretable knowledge. This makes them perfect for modelling documents' content. To create knowledge graphs of records, document-level feature extraction that captures relationships through vast sentences can be used. These approaches concentrate on removing all persons and relationships from a record that can be hundreds or thousands of long and dense documents, such as research articles [12, 24].

The automated summary that tackles recognizing salient details in a text and presents the discourse structuring's as an additional challenge [15, 25]. The direct summary of individuals and relationships as a first step could decouple the mixed pressures on models and help ensure the factual accuracy of a summary in consistent with current trends in the summary assessment [7, 23, 26].

Most of the data extraction research focuses on extracting individuals and their relational information from a single sentence [20, 26]. The latest research discusses document-level information extraction aimed at capturing relationships across distant texts or filling out a pre-defined schema table with entities [10].

The task is formulated as the classification of the relationship sort between each pair of bodies of ground truth represented in the paper. Do not believe that ground-truth objects exist and derive relationships at the document level directly from the text. This paper has created a subject knowledge base and proposed a description of an unsupervised Chinese conceptual parsing architecture. These approaches are examples of the few research in which external information is incorporated into unsupervised summary models [9].

Our work enhances this trend of applying summarization to the comprehension of long documents while addressing the need for oriented, concise representations of information. The concept of using weak monitoring to classify salient individuals linked to experimental outcomes is discussed in the research [10]. Few studies have centered on unsupervised learning approaches for document summarization, such as Text rank [27].

Document topics are often valuable knowledge, and often researchers concentrate on the impact of topics to produce better quality summaries. Real-world information graph entities are also meaningful knowledge in addition to subjects that can boost the consistency of summaries [13]. While our suggested concept of summarizing and producing summarized information graphs can be extended in any field to documents, concentrate on scientific articles throughout this work. Current work to explain scientific records involves and is therefore not limited to the extraction of knowledge [22].

4   Research Methodology

This research study focuses on the application of knowledge graph and summarization of candidates' profiles. The approach used for summarization and generating knowledge graph is discussed in this section in details.

4.1   Summarization Approach

In the current scenario, most IT companies get soft copies of candidates' profiles for any open position. The format of the application varies and comes from different sources. This huge amount of information from various sources and in different formats needs to be summarized effectively for optimum usage. To do that, we need to use the summarization techniques on the candidates' profile corpus. There are two types of summarization approaches in text analysis:

1. Extractive Summarization
2. Abstractive Summarization

4.1.1   Extractive Summarization

In the extractive summarization approach, the exact sentence from the original text content is extracted based on the key segment's importance. There are different approaches for extractive summarization out which few are given below:

1) Topic Words Based Approach

2) Frequency Based Approach

    a.   Word Probability based approach
    b.   TF-IDF based approach

In this study, page rank algorithm is used for summarization. There are various steps to implement the summarization algorithm on text data as below:

Step 1: Get the raw corpus of text for summarization.
Step 2: Remove stop words from the corpus and do POS tagging as noun and adjective are mostly important for summarization.
Step 3: Break each sentence in words.
Step 4: Tokenize the sentence and words in sentences.
Step 5: Calculate TF-IDF and similarity between sentences.
Step 6: Apply page rank algorithm to score the sentences for summarization.
Step 7: Select the top rank as per score from page rank to form the summary. In general, people use the top five sentences.
Step 8: Compare the result using the Rouge score.

All the above steps are followed for summarization of candidates' profile using page rank algorithm to rank the sentences in the candidate profile. The process flow of the analysis is shown in Fig. 1 below:

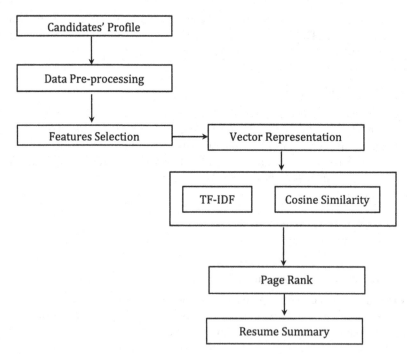

Fig. 1. A process for candidate profile Summarisation using Pagerank algorithm

4.1.2 Abstractive Summarization

It is an NLP-based approach, in which it first examines and interprets the actual context of the entire text and produces the summary. It needs high computing power and uses done through deep learning and neural network. Most of the summarization done by human is not classified as extractive, but most of the research in progress is around extractive summarization only. It is also observed that the extractive summarization gives better results than the abstractive summarization [8].

4.1.3 Statistics to Evaluate the Summarization Output

Summarization is not a simple task and measuring the accuracy of the summary is not easy. Human summarizers have been found to have poor consensus in assessing and generating summaries. Furthermore, the extensive use of different variables and the lack of a common measurement criterion have made it challenging and daunting to summarize the assessment discussed by [19].

Recall Oriented Understudy for Gisting Evaluation (i.e., Rouge) is measured to determine the summary's efficiency and accuracy. Another research implemented it to automatically decide the precision of explanation by comparing it with human reference summaries [14]. There are other variants of ROUGE as discussed below:

ROUGE-n: This metric is a recall-dependent measure and a sequence of n-grams (most frequently two, three and sometimes four) and is based on a comparison of n-grams from the reference summary and candidate summary.

Let 'm' indicates the number of common n-grams taken from the reference summary between the nominee and the reference summary, and n indicates the number of common n-grams taken from the reference summary. To calculate the score, the below formula is used

$$\text{ROUGE-n} = m/n. \tag{1}$$

ROUGE-L: This approximation uses the concept of the longest common subsequence.

ROUGE-SU: All bi-grams and unigrams are considered by this test, called Skip Bi-gram and ROUGE unigram. This measure allows terms to be inserted between the bi grammes' first and last terms, so sequences of words do not need to be consecutive.

4.2 Knowledge Graph Visualization

Some of the most common methods used, such as conceptual approaches to visualization based on frequency methods, graph intelligence and machine learning techniques [2]. However, it is not easy to comprehensively describe all the various algorithms and approaches. This study aims to provide a clear understanding of advanced developments in automated methods of summary and visualization to assist in the recruitment process.

The technique for visualizing the knowledge graph for the profiles of candidates is to present all relevant and related information in the profile. The knowledge graph is nothing but a set of connected nodes or entities with a defined relationship as edges. One can have multiple relations existing between multiple nodes in the case of a large text corpus. For example, several candidates' profiles may have the same skill sets and experience. They may have also worked in the same organization in the past. Such a relationship can easily be visible with the help of a knowledge graph in Fig. 2 below:

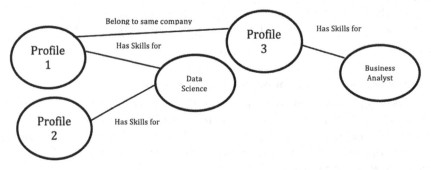

Fig. 2. Knowledge graph with multiple relationships

5   Data Analysis

The data analysis is conducted on the sample data source from an IT company. During this analysis the first step was to be pre-processing the candidates' profiles using natural language processing for summarization and knowledge graph generation.

5.1   Population and Sample of the Study

To conduct this research study, candidates' profiles were sourced from one of the mid-sized IT organizations. The data was shared by the IT organisation's human resource department, in different formats (i.e., pdf, doc, docx, etc.). The distribution of the different formats of the source data (candidates' profiles) is shown in Table 1 below:

Table 1. Knowledge graph with multiple relationships

Profile format	# of profile
MS Word (*.doc and *docx)	686
PDF Format	273
Other Formats	70

The distribution above shows that most of the profiles were sourced in Microsoft word document format. The study can be extended on a large amount of data in future.

5.2 Candidates' Profile Summarization

In the process diagram, the candidate profile is first pre-processed with initial NLP steps and then the similarity is calculated. Once the similarity between sentences is known, the summarization of the candidate's profile is executed by using page rank algorithms.

The raw text corpus prepared initially in the study for summarization. The data pre-processing steps were applied to the individual candidate's profile. The output of data pre-processing for one candidate's profile is shown in the Fig. 3 below:

```
●●●                         📄 corpus.txt — Edited
Avyan PROFESSIONAL SKILLS SUMMARY Hands on experience in large data distributed processing
using Hadoop mapreduce, pig and hive on Linux OS. Testing with MRUint Proficient working
and developing dataload jobs using ETL tools such as Talend Integration and Big Data
Seasoned professional in software development using Java. Following steps of Agile testing
with JUnit Expert in application of Multivariate Statistics and Time Series Analysis.
Hands on Experience in SAS  Base Macro Stat Enterprise Miner and R. Dedicated hard working
individual with excellent inter communication skill to network with all levels of the
organization. Proven team leader and negotiator with managing meetings with senior level
officers. EDUCATIONAL QUALIFICATION Bachelor of Engineering in Mechanical Engineering from
Utkal University Orissa passed with First Class Grading Bachelor of Science with Chemistry
at BJB College Bhubaneswar Orissa. EMPLOYERS AND POSITIONS JPS ASSOCIATES PRIVATE LIMITED
New Delhi
```

Fig. 3. Distribution of profile formats

Once the data is ready to implement the algorithm, we need to choose an appropriate summarization approach. There are several approaches for summarization, as discussed in the previous chapter. We have applied the "text rank" algorithm to summarize the text corpus. Text rank algorithm produced the most significant result (Source). The output of the text rank summarizer is shown in Fig. 4 below.

```
Avyan PROFESSIONAL SKILLS SUMMARY Hands on experience in large data distributed processing using Hadoop mapreduce, pi
g and hive on Linux OS.Testing with MRUint Proficient working and developing dataload jobs using ETL tools such as Ta
lend Integration and Big Data Seasoned professional in software development using Java.Following steps of Agile testi
ng with JUnit Expert in application of Multivariate Statistics and Time Series Analysis.
```

Fig. 4. Summarization output for a candidate's profile

The above Fig. 4 is generated by the summarization algorithm. To check the quality of the summary or coverage of the original text's actual context, we calculated ROUGE score. The ROUGE score will help understand how many words in the exact text or human reference summary appeared in the summarization algorithm's output. The ROUGE score is shown in Fig. 5 below:

```
ROUGE-1: 0.6206896551724138
ROUGE-2: 0.6153846153846153
ROUGE-L: 0.6206896551724138
ROUGE-BE: 0.6896551724137931
```

Fig. 5. Rouge score generated for candidate profile summary

If many words from the original candidate's profile text appearing in the algorithm results, it shows high ROUGE. In this study, the ROUGE score is more than 0.60, which means that 60% of the keywords appear in the summary of the actual text. The summarization result will help the recruiter look at the summary of the candidate's profile quickly.

This summary is generated from the original text. There is no "right" summary of a text. One has to decide on a metric or combined metric for scoring sentences. For example, use a word frequency-based metric (as the starter code is set up to do), or a metric based on a sentence's position in the text, or some other relevant criterion.

5.3 Knowledge Graph Generation for Candidates' Profiles

This section generates the summarization and visualization of the candidates' profiles corpus to give a quick understanding of the candidates' profiles. Knowledge graph visualization for candidates' profiles will help to reduce the effort and time to look at the profile characteristics. It will also help the recruiters to screen the applicants quickly.

For data analysis purposes and better visualization, only three profiles are considered for analysis and it can be generalized for more. The candidates' profile corpus to generate the knowledge graph it needs to split the profiles into sentences as per the process. To perform this analysis, the python package "Spacy" is used.

The sentence segmentation is done for three candidates' profiles, which identifies the key entities' pair in the candidates' profiles. The key entities' pair extracted from the above sentence segmentations are shown in Fig. 6 below:

```
[['Avyan', 'College Bhubaneswar Orissa'],
 ['Avyan', 'mapreduce'],
 ['Ruby', 'Data scientist'],
 ['Ruby', 'R'],
 ['Ruby', 'python python'],
 ['Sida', 'Stat Enterprise Miner'],
 ['professional software Avyan', 'Java Development'],
 ['Avyan', 'senior level officers'],
 ['bachlor', 'statistics'],
 ['Avyan', 'Mechanical Engineering'],
 ['Avyan', 'Agile Multivariate Statistics'],
 ['Sida', 'Agile Multivariate Statistics'],
 ['Avyan', 'Stat Enterprise Miner'],
 ['Sida', 'deep car damage learning'],
 ['Proficient', 'such Talend Integration'],
 ['Ruby', 'working Machine python'],
 ['Ruby', 'online Data R'],
 ['bachlor', 'statistics'],
 ['Ruby', 'data science']]
```

Fig. 6. Key entities pair in candidates' profile

These key entities' pairs will be used to extract the relationship among the subject and object as shown below in the Fig. 7:

```
has              6
is               3
done             3
got              2
worked on        2
completed        1
is Proficient    1
built            1
is confident     1
dtype: int64
```

Fig. 7. Relations found in candidates' profile

In Fig. 7 above, it has several relationships between the different subjects and objects in the sentences. In the next step, the extracted source and target from each sentence of the candidates' profiles is shown in Table 2 below:

Table 2. Subject, predicate and object identified in the candidates' profile corpus

	source	target	edge
0	Data I	MS excel	completed
1	Avyan	College Bhubaneswar Orissa	got
2	Avyan	mapreduce	has
3	Ruby	Data scientist	is confident
4	Ruby	R	has

The extracted source and target along with the relationship., in other words, edge in the sentence. For example, the source, "Avyan" got a target "MapReduce" with an edge "has". This can convey that the candidate has the skill set on MapReduce in his/her profile.

Now, this relationship table along with source and target will be used for visualization. The visualization is done by using the python library called "Networkx". This network graph library helps in exploring the network graph and identifies the pattern. Figure 8 below shows the knowledge graph of the three candidates' profiles:

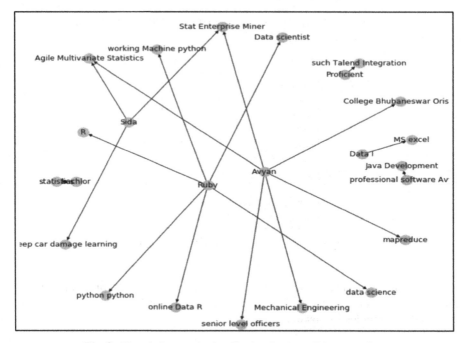

Fig. 8. Knowledge graph visualization for 3 candidates' profiles

The graph clearly shows the common skills occurring in different profiles. For example, Agile and Multivariate Statistics appear in the profiles of two candidates. Similarly, two nodes are connected with Stat and Enterprise Miner node. This helps the recruiter review and target the specific profile based on the required skills set visually. There are few nodes in the knowledge graph, that are not connected with any candidate's node. This means that there is no clear relationship established between candidates' nodes and the skillset.

6   Conclusion

The study results show that the generated summary is almost covering the candidate's full profile with high ROUGE score. NLP and extractive summarization techniques can help recruiters get a quick summary of the candidates' profiles. This approach will help recruiters process many profiles and get a quick view instead of a full profile screening. The Knowledge Graph extraction and visualization give insights to review the individual applicants or group applicants' profiles. In the analysis, extracted triplet, i.e., subject, relationship, and object are very useful for generation of knowledge graph. The visualization of the triplet is done using a network graph that shows the profile characteristics and relationship. This will help recruiters link different applicants' profiles and analyze how profile attributes are linked.

7 Discussion and Future Research Direction

The next idea is to strengthen the application of analytics in recruitment process. As technologies are evolving, researchers can explore advance machine learning and deep learning algorithms such as natural language generation and abstractive summarization for robust solution. The study also suggests a lot of real-time insights into the recruitment process generated by these new analytical-based recruitment methods. There are opportunities to enhance this insight through more research in this area. According to the related research literature, many machine learning techniques can be used together to produce a better approach for the knowledge graph generation. Most of the recruiter is using the traditional approaches for mining the relevant information from candidates' profile, the study proposed a new approach to extract, summarize and visualize the information.

References

1. Abuobieda, A., Salim, N., Kumar, Y.J., Osman, A.H.: An improved evolutionary algorithm for extractive text summarization. In: Selamat, A., Nguyen, N.T., Haron, H. (eds.) ACIIDS 2013. LNCS (LNAI), vol. 7803, pp. 78–89. Springer, Heidelberg (2013). https://doi.org/10.1007/978-3-642-36543-0_9
2. Allahyari, M., Safaie, S., Pouriyeh, S., Trippe, E., Assefi, M., Gutierrez, J.: Text summarization techniques: a brief survey. Int. J. Adv. Comput. Sci. Appl. 8(10), 397–405 (2017)
3. Bharti, S., Babu, K., Jena, S.: Automatic keyword extraction for text summarization: a survey. Eur. J. Adv. Eng. Technol. 4(6), 410–427 (2017)
4. Bondarouk, T., Parry, E., Furtmueller, E.: Electronic HRM: four decades of research on adoption and consequences. Int. J. Hum. Resour. Manag. 28(1), 98–131 (2017)
5. Canhasi, E.: Graph-based models for multi-document summarization. A Dissertation: Faculty of Computer and Information Science. http://lkm.fri.uni-lj.si/xaigor/eng/scipaper/AASummarization.pdf. Accessed 1 Dec 2020
6. Devihosur, P., Naseer, R.: Automatic text summarization using natural language processing. Int. Res. J. Eng. Technol. 4(8), 1434–1440 (2017)
7. Durmus, E., He, H., Diab, M.: FEQA: a question answering evaluation framework for faithfulness assessment in abstractive summarization. In: Proceedings of the 58th Annual Conference of the Association for Computational Linguistics (ACL), pp. 5055–5070 (2020). https://doi.org/10.18653/v1/2020.acl-main.454
8. Günes, E., Dragomir, R.: LexRank-graph-based lexical centrality as salience in text summarisation. J. Artif. Intell. Res. (JAIR) 22(1), 457–479 (2004)
9. Hou, S., Lu, R.: Knowledge-guided unsupervised rhetorical parsing for text summarization. Inf. Syst. 9(9), 1520 (2020). https://doi.org/10.3390/electronics9091520
10. Jain, S., Zuylen, M.V., Hajishirzi, H., Beltagy, I.: SciREX: a challenge dataset for document-level information extraction. In: ACL (2020). https://ui.adsabs.harvard.edu/abs/2020arXiv200500512J
11. Jia, J., Miratrix, L., Gawalt, B., Yu, B., Ghaoui, L.: Summarizing large-scale, multiple-document news data: sparse methods and human validation. Annals of Applied Statistics. https://statistics.berkeley.edu/sites/default/files/tech-reports/801_0.pdf. Accessed 20 Nov 2020
12. Jia, R., Wong, C., Poon, H.: Document-level N-array relation extraction with multiscale representation learning. In: NAACL-HLT. arXiv abs/1904.02347, pp. 3693–3704 (2019)

13. Kim, S.E., Kaibalina, N., Park, S.B.: A topical category-aware neural text summarizer. Appl. Sci. **10**(16), 5422 (2020). https://doi.org/10.3390/app10165422

14. Lin, C.: Rouge: a package for automatic evaluation of summaries. In: Text Summarization Branches Out Proceedings of the ACL-04 Workshop, vol. 4, pp. 74–81 (2004)

15. Liu, Y., Lapata, M.: Text summarization with pretrained encoders. In: EMNLP (2019)

16. Lloret, E., Palomar, M.: COMPENDIUM: a text summarization tool for generating summaries of multiple purposes, domains, and genres. Nat. Lang. Eng. **19**(2), 147–186 (2012)

17. Ozsoy, M., Alpaslan, F.: Text summarization using latent semantic analysis. J. Inf. Sci. **37**(4), 405–417 (2011)

18. Radev, D., Budzikowska, H., Budzikowska, M.: Centroid-based summarization of multiple documents. Inf. Process. Manag. **40**(6), 919–938 (2004)

19. Saggion, H., Thierry P.: Automatic text summarization: past, present and future. In: In: Poibeau, T., Saggion, H., Piskorski, J., Yangarber, R. (eds.) Multi-source, Multilingual Information Extraction and Summarization, pp. 3–13. Springer, Heidelberg (2016). https://doi.org/10.1007/978-3-642-28569-1_1

20. Stanovsky, G., Michael, J., Zettlemoyer, L., Dagan, I.: Supervised open information extraction. In: NAACL (2018). https://doi.org/10.18653/v1/N18-1081

21. Tandon, L., Joshi, P., Rastogi, R.: Understanding the scope of artificial intelligence in human resource management processes–a theoretical perspective. Int. J. Bus. Adm. Res. Rev. **1**(19), 62–67 (2017)

22. Wadden, D., Wennberg, U., Luan, Y., Hajishirzi, H.: Entity, relation, and event extraction with contextualized span representations. In: EMNLP, pp. 5784–5789 (2020)

23. Wang, A., Cho, K., Lewis, M.: Asking and answering questions to evaluate the factual consistency of summaries. In: ACL (2020)

24. Yao, Y., et al.: DocRED: a LargeScale document-level relation extraction dataset. In: ACL, pp. 764–777 (2019). https://doi.org/10.18653/v1/P19-1074

25. Yasunaga, M., et al.: ScisummNet: a large annotated corpus and content-impact models for scientific paper summarization with citation networks. In: Proceedings of the AAAI Conference on Artificial Intelligence, vol. 33, no. 1, pp. 7386–7393 (2019). https://doi.org/10.1609/aaai.v33i01.33017386

26. Zhang, Y., Zhong, V., Chen, D., Angeli, G., Manning, C.D.: Position-aware attention and supervised data improve slot filling. In: EMNLP, pp. 35–45 (2017). https://doi.org/10.18653/v1/D17-1004

27. Zheng, H., Lapata, M.: Sentence centrality revisited for unsupervised summarization. arXiv 2019, arXiv:1906.03508

Knowledge and Innovation Management for Transforming the Field of Renewable Energy

Olena Trofymenko[1]($\boxtimes$) ⓘ, Olena Shevchuk[2] ⓘ, Nataliia Koba[3] ⓘ,
Yurii Tashcheiev[4] ⓘ, and Tetiana Pavlenco[5] ⓘ

[1] National Technical University of Ukraine "Igor Sikorsky Kyiv Polytechnic Institute",
International University of Finance, 37, Prosp. Peremohy, Kyiv 03056, Ukraine
o.o.trofymenko@gmail.com
[2] National Technical University of Ukraine "Igor Sikorsky Kyiv Polytechnic Institute",
37, Prosp. Peremohy, Kyiv 03056, Ukraine
shevchuk-oa@ukr.net
[3] International University of Finance, Kyiv, Ukraine
ashatunka@gmail.com
[4] Institute of Renewable Energy of the National Academy the Sciences of Ukraine, 20a,
Hnata Khotkevycha Str., Kyiv 02094, Ukraine
tascheevyuri@gmail.com
[5] National Technical University of Ukraine "Igor Sikorsky Kyiv Polytechnic Institute",
37, Prosp. Peremohy, Kyiv 03056, Ukraine
pavlenco.tetiana@lll.kpi.ua

Abstract. The article examines the features of renewable energy development at the macro and micro levels in Ukraine. The role of knowledge and innovation management in the field of renewable energy has been defined. **The subject** of the research is the processes of knowledge and innovation management at renewable energy enterprises. The purpose of the article is to identify key factors in the development of renewable energy and approaches to knowledge and innovation management at renewable energy enterprises. General scientific methods have been used, the main of which are: system analysis – application of a comprehensive two-level approach to the analysis of the role of knowledge in the field of renewable energy and correlation and regression analysis. The following **results** have been obtained: based on the conducted analysis, the expediency of forming an innovation infrastructure for developing such a high-tech sector as renewable energy has been justified. The application of methods and tools of knowledge and innovation management at renewable energy enterprises has been substantiated. Recommendations concerning the formation and development of a knowledge and innovation management system at renewable energy enterprises in Ukraine have been made, as this system will increase the competitiveness of renewable energy in relation to traditional one. **Conclusions:** the role and features of knowledge and innovation management in renewable energy enterprises are shown. Prospects for the development of renewable energy in Ukraine on the basis of knowledge and innovation management are identified.

Keywords: Knowledge management · Renewable energy · Innovations

© Springer Nature Switzerland AG 2021
A. Solanki et al. (Eds.): AIS2C2 2021, CCIS 1434, pp. 73–87, 2021.
https://doi.org/10.1007/978-3-030-82322-1_6

1 Introduction

The development of renewable energy (RE) is one of the priority tasks in accordance with the adopted Energy Strategy of Ukraine for the period up to 2035 year "Security, energy efficiency, competitiveness" [1]. The goal 7 "Affordable and Clean Energy" of the Global Sustainable Development Goals by 2030, approved at the UN Summit on Sustainable Development [2] also includes this task. Also, there is an agreement with the UNFCCC [3], including in part of the implementation of projects aimed at environmental protection. All this contributes to the growth of RE projects and increase the number of enterprises engaged in their implementation in Ukraine.

It is well known that the field of RE is a high-tech industry, so to ensure its development it is necessary to implement effective approaches to knowledge and innovation management. In such a case, it is important to analyse the sectoral characteristics, indicators of RE in Ukraine in order to determine the main factors in the knowledge economy that will affect its development. In particular, it is essential to investigate the impact of elements of the knowledge economy on the development of RE and the role of knowledge and innovation management in the formation and development of RE; to analyse and identify modern approaches to knowledge and innovation management conditioned by sectoral characteristics; to establish priority areas of the development of RE on the basis of knowledge and innovation management.

The scientific novelty of the paper consists in the fact that a comprehensive approach to analysing knowledge and innovation management in the field of RE in Ukraine was applied; an analysis of factors and approaches to knowledge management at two levels - macro and micro levels - was conducted. In addition, while studying the features of knowledge and innovation management at the level of RE enterprises, we applied the conceptual provisions of knowledge management in an organisation on the basis of scientific approaches, which, unlike existing ones, are based on the use of a value chain model, which provides organisational and methodological support for building, development and effective use of intellectual capital in the growth of value added.

2 Literature Review

One of the key factors in the RE companies development is their willingness to innovate. Under these conditions, the formation of a knowledge management system becomes especially important [4, 5], which involves a systematic process of transforming the experience of employees into processes and results that increase the total cost and overall productivity.

As Hayek [6] rightly pointed out back in 1945, every economic entity has unique knowledge. Later Penrose [7] in his research came to the conclusion that any enterprise is a repository of knowledge that characterizes it as an administrative organization and as a set of material and human resources. Marshall in his works [8] generally attributed knowledge to the capital of the company, claiming that it is the driving force of the production. Nelson & Winter [9] note that the basis of the evolutionary development of the enterprise is a natural way of technological changes, called by specific knowledge, unique to each enterprise. Due to these assumptions, this study will regard RE companies as those that have a unique set of knowledge that can determine their technological development and success in the marketplace.

Obviously, to ensure such success is not enough just to have a unique set of knowledge, you need to be able to manage it as capital. In the work [10], the authors emphasize the importance of using organizational education to adapt to changes arising from technological and economic innovations. Other researchers [11] focus on the elements of the knowledge transfer process - human capital, tasks that reflect the goals and mission of the enterprise and tools in the form of hardware, software and technology. Finally, Drucker [12] argues that the main problem facing any company is to create a systematic method of knowledge management, which is to continuously improve existing and search for new activities through innovation. In summary, the authors focus their research on individual processes related to the movement of knowledge in the organization (knowledge creation, learning, knowledge transfer, storage), however, as Edwards [13] rightly points out, there is a need for holistic, systemic, integrated approaches to deal with the ever-increasing complexity of organizations in knowledge management in enterprises. Thus, in this paper, it is proposed to consider both the activities of RE companies and knowledge management within these companies from the standpoint of a systems approach. Therefore, knowledge management will be viewed by us as a set of purposeful, systematic and organized management actions for processes in the field of knowledge using various tools and technologies that are aimed at achieving goals and increasing the competitiveness of the organization.

Knowledge management, like any management system, involves the presence of a subject and object of management. In this study such a subject is the processes in the field of knowledge. This statement is fully consistent with the approach that knowledge requires certain actions to update it in the enterprise. In addition, according to the theory of dynamic capabilities [14], enterprises find new ways for dynamic development by recombining existing capabilities, which are determined by existing knowledge. That is, knowledge should be considered in the dynamics, in constant motion from the emergence at the employee level to the use of the organization. In the scientific literature, such processes in the field of knowledge are called the process of knowledge transformation, which Nonaka and Takeuchi [4] consider the main source of competitiveness.

We agree with Wier et al. [15] that for modern enterprises, in particular, enterprises in the field of RE, the highly significant knowledge management processes are considered to be internal, i.e. knowledge sharing within the organization, but also a knowledge flow from external agents has emerged as important. Therefore, it is necessary to consider the transformation processes of knowledge in RE enterprises in close connection with the external environment [39]. This relationship reflects the interaction of all types of knowledge - organizational, individual and relative.

As Edwards [13] aptly points out, although a few aspects of knowledge management in the energy sector are sector-specific, noticeably the need to manage knowledge on relatively long timescales, most issues are common to organizations. That is why the scheme of knowledge transformation in organizations will be used to analyse the practical experience of RE companies and identify features of knowledge management in this area.

3 Methodology

The field of RE is developing rapidly and is a high-tech and innovative industry. Any innovation and technology is commercialized knowledge, thus, to increase the efficiency of RE companies, it is important to explore the factors that affect the process of knowledge transformation.

That is why in our study we have used a two-level approach to the analysis of the role of knowledge for RE, namely: 1) the macro level – the identification of sectoral features and the impact of the knowledge economy factors on the development of RE in Ukraine. 2) the micro level – the research and establishment of specific features of knowledge management processes at RE enterprises, in particular, the role of the learning factor in the cost of RE technologies.

To study the impact of the factors on RE development, we have chosen the resulting indicator – the capacity of RE, and such factors of the likely impact as: the Global Innovation Index (GII) – a generalized measure of innovation in the country, which contains important for our study sub-indices like Knowledge & technology outputs, Human capital & research and Creative outputs, which testify to the state and development of the knowledge economy at the macro level; the number of licensees, who received a "green tariff".

To determine the closeness of the correlation between the selected variables we have used a paired linear correlation-regression model [16]:

$$\tilde{y} = b_0 + b_1 x \tag{1}$$

where $\tilde{y}$ is the theoretical (regulative, forecast) value of the resulting variable y; b_1 is regression coefficient, tangent of the angle of inclination of the straight line; b_0 is the free term of the regression equation, the initial value of the resulting variable y is the resulting variable, the capacity of RE; x is a factorial feature, the indicator of the global innovation index for the first definition of the first connection and the number of licensees, who received a "green tariff" for the second connection.

To determine the b_0 and b_1 unknown correlation parameters β_0 and β_1, the least squares method, as one of the most effective methods for estimating the parameters of econometric models, was used. The calculation of coefficients of change in the capacity of RE taking into account the influence of GII as an indicator of the knowledge economy development at the macro level and the number of licensors who received a "green tariff" as an incentive indicator for introducing innovations at RE enterprises will be carried out according to the system of normal equations [42]:

$$\begin{cases} nb_0 + \left(\sum_{i=1}^{n} x_i\right)b_1 = \sum_{i=1}^{n} y_i \\ \left(\sum_{i=1}^{n} x_i\right)b_0 + \left(\sum_{i=1}^{n} x_i^2\right)b_1 = \sum_{i=1}^{n} x_i y_i \end{cases} \tag{2}$$

where n is the number of periods; x is indicators of the global innovation index and separately the number of licenses; y is the capacity indicator.

The pairwise correlation coefficient r has been calculated by the formula [42]:

$$b_1 = \frac{cov_{xy}}{D_x} \tag{3}$$

where D_x is the variance of the random variable x, measured in square units of the variable x.

$$r = \frac{\sum_{i=1}^{n}(x_i - \bar{x})(y_i - \bar{y})}{\sqrt{\sum_{i=1}^{n}(x_i - \bar{x})^2} \cdot \sqrt{\sum_{i=1}^{n}(y - \bar{y})^2}} \qquad (4)$$

The equation of the direct line passing through the point $(\bar{x}, \bar{y})$ parallel to the indirect vector $V = \left(\frac{S_x}{r}, S_y\right)$ is:

$$\frac{\tilde{y} - \bar{y}}{S_y} = r\frac{(x - \bar{x})}{S_x} \qquad (5)$$

where $Sx = \sqrt{D_x}$ is the sample standard deviation (the standard) of the variable x,
$Sy = \sqrt{D_y}$ is the sample standard deviation (the standard) of the variable.
The coefficient of determination is an indicator of adequacy for the applied model. It is calculated by the formula:

$$R^2 = 1 - \frac{\sum (y - y^*)^2}{\sum (y - y_{cep})^2} \qquad (6)$$

where y^* is the calculated predicted values of y.
To determine the accuracy of the calculations, we calculated the average approximation error:

$$A = \frac{\sum \frac{|y - y^*|}{y^*}}{n} \times 100\% \qquad (7)$$

At the micro level, the main features of knowledge management and innovation in the field of RE were analysed, taking into account the theoretical provisions and analysis of practical experience of domestic and foreign countries in accordance with the process of knowledge transformation. To develop recommendations for building an effective knowledge management system at a particular RE company of a certain type of energy (solar, wind, bioenergy, etc.), the article suggests applying Porter's value chain [41], which helps to identify unique business processes and knowledge involved in them and necessary innovations. In our work, using the value chain of solar and wind energy enterprises, we identified the knowledge involved in the implementation of business processes and the stages of its transformation, which are reflected by the chain of intellectual support of the process of creating consumer value.

4   Knowledge Economy as a Factor of the RE Development

Today, the RE sector in Ukraine is developing. In total, in 2019 the production of electricity by alternative sources (wind farms, solar power plants, biomass) amounted to 5542.2 million kWh, which is 2909.5 million kWh, or 110.5% more than in 2018. From 2015 to the first quarter of 2020, the capacity of renewable electricity facilities (excluding the temporarily occupied territory of the Autonomous Republic of Crimea), which has a "green tariff", increased by 6,727 MW, of them put into operation: in 2015 - 32 MW; in 2016 - 136 MW; in 2017 - 291 MW; in 2018 - 848 MW; in 2019 - 4,658 MW; in the first quarter of 2020 - 763 MW [17]. Today in Ukraine there are more than 380 enterprises [18] in the field of RE and energy efficiency, which, together with the results of the industry as a whole, indicates the importance of providing comprehensive support for their development.

Table 1. Data for estimating the impact of the Global Innovation Index and the number of licensees who received a "green tariff" on the capacity of RE in Ukraine.

Year	Capacity, MW Y_{CAP}	The Global innovation index X_{GII}	Capacity (projected) Y^*_{CAP1}	Number of licensees X_{num}	Capacity (projected) Y^*_{CAP2}
2013	5769	35,8	5881	107	5823
2014	6048	36,3	6140	145	6071
2015	6105	36,5	6243	131	5979
2016	6199	35,7	5830	163	6188
2017	6530	37,6	6811	230	6626
2018	7530	38,5	7276	363	7494
Sum	38181	220,4		1139	
r		0,9136		0,9923	
R^2		0,8347		0,9846	
$\bar{A}$		3,23		0,91	

Since energy, in particular, RE is a high-tech science-intensive field of activity, which is designed to serve all areas of activity and meet the needs of society, it is important to ensure innovative development of the country and implement modern approaches to knowledge management [36]. Today, in the time of the knowledge economy development and the formation of the foundations of Industry 4.0, it is important to investigate the impact of innovation, as an infrastructure for the high-tech industry, on the indicators of RE.

The GII [19] was chosen as such a factor, which indicate the state and development of knowledge economy at the macro level. At the same time, one of the main result indicators of the development of RE is capacity [40]. Therefore, it was chosen as the resulting variable. It is also interesting to determine the impact of such a factor as the number of licensees who received a "green tariff" on the capacity of RE. The input data and calculation data for finding the correlation between the factors are presented in Table 1.

To construct mathematical dependence, we solved a system of equations to determine the influence of each factor on the capacity of RE. In particular, the systems of Eqs. 8 and 9 for the GII and the number of licensees are given below.

$$\begin{cases} 6b_0 + b_1 \times 220, 4 = 38181 \\ b_1 \times 220, 4 + b_0 \times 8102, 08 = 1405642, 4 \end{cases} \tag{8}$$

$$\begin{cases} 6b_0 + b_1 \times 1139 = 38181 \\ b_1 \times 1139 + b_0 \times 260873 = 7539725 \end{cases} \tag{9}$$

where b_0, b_1 – are linear regression coefficients.

According to the calculations, the linear models were obtained (Fig. 1, Fig. 2):

– for the global innovation index (GII):

$$y_{CAP} = -12611, 84 + 516, 57x_{GII} \tag{10}$$

– for the number of licensees who received a "green tariff":

$$Y_{CAP} = 5123, 89 + 6, 53x_{num} \tag{11}$$

Figure 1 shows the dependence of the capacity of RE on the GII. The correlation value is 0.91, which indicates a close relationship between GII and capacity. It should be noted that with an increase in the value of GII by 1 point the capacity is expected to increase by 516,75 MW in accordance with the linear regression.

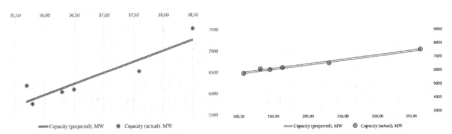

Fig. 1. Dependence of RE capacity on the GII *Source: calculated by the authors*

Fig. 2. Dependence of RE capacity on the number of licensees *Source: calculated by the authors*

Additionally, we compared the capacity with the number of licensees who received a green tariff from 2013 to 2018 (Fig. 2). We obtained a close relationship with a correlation coefficient of 0.99, which indicates a close relationship between the variables. The number of licensed enterprises is constantly growing. Compared to 2013, the number of licensors has more than doubled, all this indicates that this market segment is not monopolized (weakly concentrated) and the growth of enterprises will increase capacity. It also indicates that with the development of their activities, companies will need to implement approaches to knowledge management in the corporate governance system.

One of the crucial factors in the development of RE is the cost component of the equipment used for the installation of regenerative systems (RES systems). In turn, forecasting the cost of these technologies very often consists of a number of interrelated mathematical models, in which there is a learning factor, which is one of the components of knowledge management. In our opinion, in the context of predicting the cost of RES technologies and using the learning ability indicator, the S-curve model of technology development, combined with the Wright cost model [20] is the most interesting [37].

The S-curve model of technology development is used to describe innovative technologies when entering the market, linking the time and penetration of innovation to the market by functional dependence and is a logistic equation:

$$f'(t) = k \times f(t) \times (S - f(t)) \tag{12}$$

where f (t) – quantity at time t; f'(t) – the rate of change in time t: k – growth factor; t – time moment; S – saturation limit (t).

The solution to the equation is the logistic function, S-curve, (logistic curve):

$$f(t) = \frac{f(0)}{f(0) + (S - f(0)) \times e^{-Skt}} \tag{13}$$

where $f(0)$ – quantity at time $t = 0$.

It may be possible to predict the dynamics of costs and prices for technology by applying an experience curve that links the experience gained as a result of manufacturing a product with the cost of this product. Concept based on learning effects, first described by Wright in 1936, in a mathematical model of aircraft manufacturing costs [20].

The concept of learning or experience is an empirical law describing the cost decline of a product in industries. Mathematically, it may be in the following form:

$$C(x_t) = C(x_0) \times \left(\frac{x_t}{x_0} \right)^{-b} \tag{14}$$

where x_t – total production corresponding to the time t; x_0 – total production corresponding to the time t_0; C (x_t) – cost (or price) at time t; C (x_0) – cost (or price) at time t_0; b – learning parameter.

$$LR = 1 - 2^b = 1 - P \tag{15}$$

where LR – learning speed.

In a double logarithmic scale, the function (14) shows a linear behaviour. The slope of this function is a measure of learning, and the steady decline in cost is described by the rate of learning (15). For the calculation, a progress rate (PR) is introduced, which is defined as one minus the learning rate. Based on this approach, the forecast of the cost of regenerative systems is carried out taking into account the learning effect. Based on the above, it may be possible to state that the rate of penetration into the market of innovative technologies, in this case, regenerative systems, depends on the level of education. In turn, it can be assumed that the level of training is functionally related to knowledge management in the field of RE.

Thus, the elements of the knowledge economy in combination with the positive dynamics of the development of RE form the favourable principles of its further development. In conjunction with the results of the industry as a whole, this indicates the importance of providing comprehensive support for their development.

Thus, the Memorandum of Understanding on the Strategic Energy Partnership between Ukraine and the EU together with the European Atomic Energy Community [21] states the need to develop Ukraine's energy infrastructure and implement projects aimed at deepening its integration into EU energy infrastructure, established principles for cooperation in attracting investment and financing.

This, together with the Energy Strategy of Ukraine, lays the foundation for the development of RE at the macro level, at the same time, the features of this area determine the needs of enterprises to build an appropriate knowledge management system for effective operation of such enterprises. After all, knowledge is a central link in the operational and strategic activities of enterprises in the RE sector and comes in various forms - scientific knowledge (for example, photovoltaics principle), technological knowledge (for example, how to effectively use wind turbines), management knowledge (for example, how to motivate employees) introduce new methods of work) [1].

The International Organization for Standardization ISO [22] has published more than 50 standards for solar energy and biofuel systems. The main areas of development of standards for RE are solar energy/power, solid biofuels, energy management and energy conservation, ecological buildings design, hydrogen technologies, automation and integration systems, industrial fans, pumps, environmental management, thermal performance and environmental energy use, capture, transportation and geological storage of carbon dioxide, information technology to ensure sustainable development, electric vehicles.

5 Transformational Processes in Knowledge Management at RE Enterprises

Generalization of the knowledge management theory states and analysis of the practical experience of companies in the field of RE allows us to identify the following features of knowledge management and innovation in this area:

1) ensuring the processes of data and information collection, formalization and storage of knowledge, as well as the transfer and diffusion of knowledge on the basis of powerful information platforms using the latest IT solutions and automated decision-making systems. The operation of RE companies is possible only on the basis of the introduction of the concept of intelligent (Smart Grid) and digital energetics. It creates opportunities for effective integration of power plants based on RE sources. Since the generation of electricity in green energy depends significantly on weather conditions, the amount of electricity produced can significantly fluctuate over time. Therefore, network management is complicated and requires an instant response, which can provide only modern IT solutions within the Smart Grid. In Ukraine, work on the introduction of new technologies in the national energy system has begun relatively recently. For example, in traditional energy, the Belgian company Tractebel

has been implementing a number of pilot technologies and Smart Grid projects since 2014 at the level of the system operator – National power company "Ukrenergo" [23]. The Ministry of Energy of Ukraine and the Korean KT Corporation signed a Memorandum of Understanding on implementation of "smart" electricity metering [24].

Ukrainian energy company DTEK is the most active in implementing technologies, installing smart meters and automating the power supply system with the help of modern software. The fastest pace of implementation of modern IT solutions is in the field of RE. This is confirmed by the fact that only 20 people are involved in servicing one of the five largest wind farms in Europe - DTEK's Botievskaya WPP. To receive the same amount of electricity at Ukrainian TPS (thermal power station) would require, on average, 12 times more staff - 225 people [25]. An example of such a powerful information platform in the knowledge management system in RE is Tethys [27]. This online platform was developed to address inaccuracies and unpredictability of environmental data in marine and wind RE [26]. The example of Tethys shows that a powerful information platform is able to provide such processes of transformation of knowledge of enterprises in the field of RE as data and information collection, formalization and storage of knowledge, as well as knowledge transfer and diffusion.

2) ensuring the processes of the transfer and diffusion of knowledge, as well as its commercialization through the introduction of new energy-saving technologies and RE sources by bringing together world-class professionals and strategic partners along the industry value chains. RE is characterized by a high level of cooperation between individual market participants. Such cooperation makes it profitable to generate electricity both in the EU and in Ukraine, even on the roofs of privately owned homes. By merging into cooperatives, for example, in Germany, rooftop solar power plant owners are able to compete in the electricity market with industrial power plants [28]. At the same time, studies of the bioenergy market in Ukraine show weak business links along value chains, the absence of long-term contracts even between biofuel producers and heat consumers [29]. A positive example of cooperation in the field of RES is the Ukrainian company Rentechno [30], which produces solar panels and integrated roofing solutions and actively cooperates with the well-known European manufacturer of photovoltaic cells and solar modules SoliTek (Lithuania). Another Ukrainian company WindFarm signed a contract with the Chinese company POWERCHINA for joint implementation of a project to build a wind farm in Ukraine [31].

3) ensuring the transfer and commercialization of knowledge through technology transfer. Thus, on the basis of the Institute of RE of the National Academy of Sciences of Ukraine [32] the International Information and Demonstration Centre for RE and Technology Transfer was established. It cooperates with well-known research universities and companies from different countries. The purpose of such a centre was the need for joint efforts to improve energy efficiency and stimulate the use of RE in various sectors of the economy, technology transfer and study, integration of new solutions in the field of RE and energy efficient technologies in Ukraine's energy infrastructure.

4) ensuring the processes of "growing" and attracting knowledge through the selection and training of highly intelligent personnel and their continuous education. According to the results of the Global Energy System report based on 100% RE, a global transition towards a 100% renewable electricity system will create over 36 million direct jobs in the power sector by 2050 [33]. It should be noted that the RES industry itself ishigh-tech and innovative, so the involved staff must be highly qualified and fully meet the criteria of a "knowledge worker". In Ukraine, an outstanding example of a responsible attitude to knowledge creation processes is DTEK Holding, which focuses on a corporate university - DTEK Academy. In addition, the corporation actively cooperates with the leading technical universities of Ukraine [25]. Thus, the holding is implementing the process of gaining knowledge by "growing" internal knowledge and involving external expertise. A positive trend for Ukraine in the field of training for energy efficiency is the introduction in 2017 of the EUREM [34]. The specifics of this program are that it provides training for professionals, who will develop real tools to save energy and reduce greenhouse gas emissions for specific enterprises [35].

5) ensuring the exchange of knowledge both within enterprises and with other market participants. In order to develop national RE enterprises, the Institute of RE of the NASU, the UWEA [38] and the National Technical University of Ukraine "Igor Sikorsky Kyiv Polytechnic Institute" hold annual international energy forums, round tables and conferences. Another example of close cooperation between RES market participants is the existence of the Ukrainian RE Association [18]. That is, the Association acts as a platform to support the external contour of knowledge transfer processes of participating organizations and promotes the development of their relative capital (component of intellectual capital, which reflects the relationship within the company and the relationship with the environment). A good example of a corporate platform for sharing knowledge within the company through the innovation community is DTEK's ID. Community, where all participants interested in new technologies communicate, learn, share expertise, and discuss existing projects and current business cases.

6) ensuring the processes of data and information collection, knowledge acquisition, its commercialization by finding and supporting innovative ideas and projects. In particular, supporting the concept of open innovations, the Ukrainian energy holding DTEK has created a platform through which anyone can submit their innovative idea through the corporation's website from anywhere in the world. At the same time, the preference is given to projects that meet the company's strategic goal - to become a carbon-neutral company by 2040 [25]. Another example of an accelerator of innovative development in the field of RES is the non-profit organization Greencubator, which implements climate, educational, social and media projects in RE. Greencubator creates an environment for meetings and development of Ukrainian "green" innovators - TeslaCamp. The organization recently became the Climate Voucher Project Manager from the EBRD and the EU, which gave it the opportunity to support Ukrainian start-ups in finding investment and drive green entrepreneurship in Ukraine.

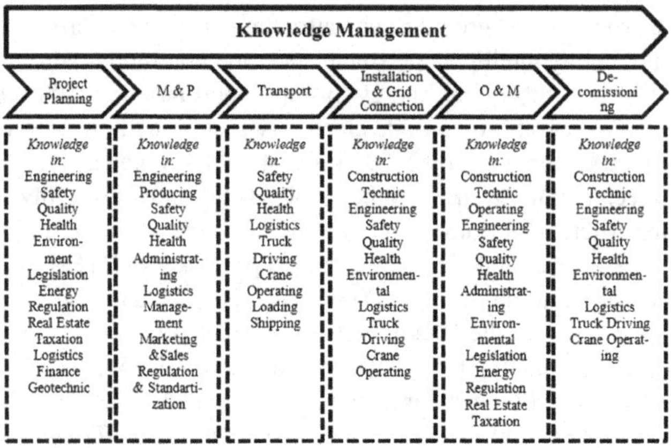

Fig. 3. The chain of intellectual support of the process of creating the value of solar and wind energy enterprise *Source: developed by the authors*

Thus, every RE enterprise to some extent has all the sub-processes of the process of knowledge transformation from its creation to commercialization. It is clear that each enterprise has a unique set of knowledge; the peculiarities of building a knowledge management system for each individual RE enterprise, for example, by type of energy (solar, wind, marine, etc.) will be manifested by this uniqueness. In our opinion, a convenient tool for determining specific knowledge is Porter's value chain [41], which makes it possible to clearly identify business processes and the knowledge involved in them, and, consequently, the necessary innovations for an enterprise. As an example, using the value chain of solar and wind energy companies, we identified the knowledge involved in the implementation of business processes and the stages of its transformation, which are reflected by the chain of intellectual support of the process of creating consumer value (Fig. 3).

It is clear that for enterprises producing biofuel, for example, this chain will be completely different and, accordingly, the set of knowledge involved will be different. Therefore, in order to build an effective knowledge management system at specific RE enterprises, these aspects should be taken into account.

6   Conclusions

The results of our comprehensive two-level study have confirmed the feasibility of developing knowledge and innovation management at the macro and micro levels in the field of RE. The correlation-regression analysis has showed a close relationship, and, therefore, high dependence of the level of capacity of RE on the GII. As the GII includes such sub-indices of knowledge development as Knowledge & technology outputs, Human capital & research, Creative outputs, and it testifies to the state and development of the knowledge economy at the macro level, it can be argued that the development of an innovative infrastructure for supporting RE is essential. In addition, the results of the analysis suggest that the market penetration index of innovative RES technologies, such as regenerative systems, depends on the level of training, which is evidenced by the model of the S-curve of technology development in conjunction with the Wright cost model.

The level of influence of the number of licencees who received a "green tariff" on the capacity of RE has also been studied and a close connection with the correlation coefficient of 0.9 has been discovered. Obtained results indicate that this market segment is not monopolized and the growth of enterprises will increase the capacity, and the introduction of knowledge management approaches will help this segment to develop further. Stimulating the development of RE enterprises will have a positive impact on the environmental situation and it will ensure a long-term positive effect due to the introduction of innovative approaches in this field. The main industry standards and regulations have been defined, as they determine the basic principles of development and regulation of the industry at the state level.

Summarizing the theoretical provisions of the theory of knowledge management and the analysis of practical experience of companies in the field of RE, we have singled out 6 main features of knowledge and innovation management in RE enterprises. We applied the conceptual provisions of knowledge management in an organisation on the basis of scientific approaches, which are based on the use of a value chain model, which provides organisational and methodological support for building, development and effective use of intellectual capital in the growth of value added.

References

1. Energy Strategy of Ukraine for the period up to 2035 "Safety, Energy Efficiency, Competitiveness": Order of the Cabinet of Ministers of Ukraine dated August 18, 2017. No. 605-p. The Verkhovna Rada of Ukraine. http://zakon2.rada.gov.ua/laws/show/605-2017-%D1%80. Accessed 10 Nov 2020
2. Sustainable Development Goals: Ukraine. Ministry of Economic Development and Trade of Ukraine. http://un.org.ua/images/SDGs_NationalReportUA_Web_1.pdf. Accessed 16 Nov 2020
3. Framework Convention "The United Nations on climate change". The Verkhovna Rada of Ukraine. https://zakon.rada.gov.ua/laws/show/995_044. Accessed 09 Nov 2020
4. Nonaka, I., Takeuchi, H.: The Knowledge-Creating Company: How Japanese Companies Create the Dynamics of Innovation. Oxford University Press, New York (1995)
5. Polanyi, M.: Personal Knowledge: Towards a Post-Critical Philosophy. University of Chicago Press, Chicago (1958)
6. Hayek, F.A.: The use of knowledge in society. Am. Econ. Rev. 35(4), 519–530 (1945). http://www.jstor.org/stable/1809376. Accessed 18 Nov 2020
7. Penrose, R.: The Emperor's New Mind: Concerning Computers, Minds, and the Laws of Physics. Oxford University Press, New York (2002)
8. Marshall, A.: Principles of Economics. An Introductory Volume. Macmillan, London (2013). https://doi.org/10.1007/978-1-137-37526-1
9. Nelson, R.R., Winter, S.G.: An Evolutionary Theory of Economic Change. Harvard University Press, Cambridge (1982)
10. Senge, P.M.: The Dance of Change: The Challenges to Sustaining Momentum in a Learning Organization, 1st edn. Currency/Doubleday, New York (1999)

11. Argotea, L., Ingramb, P.: Knowledge transfer: a basis for competitive advantage in firms. Organ. Behav. Hum. Decis. Process. **82**(1), 150–169 (2000). https://doi.org/10.1006/obhd.2000.2893

12. Drucker, P.F.: Management Challenges for the 21st Century. HarperBusiness, New York (2001)

13. Edwards, J.: Knowledge management in the energy sector: review and future directions. Int. J. Energy Sector Manag. 2 (2008). https://doi.org/10.1108/17506220810883216. https://www.researchgate.net. Accessed 18 Nov 2020

14. Kogut, B., Zander, U.: Knowledge of the firm and the evolutionary theory of the multinational corporation. J. Int. Bus. Stud. **34**, 516–529 (2003). https://www.jstor.org/stable/3557192. Accessed 18 Nov 2020

15. Weir, M., Huggins, R., Schiuma, G., Lerro, A.: Valuing knowledge assets in renewable energy SMEs: some early evidence. Electron. J. Knowl. Manag. **8**(2), 225–234 (2003). https://www.researchgate.net/publication/268328473_Valuing_Knowledge_Assets_in_Renewable_Energy_SMEs_Some_Early_Evidence. Accessed 18 Nov 2020

16. Pandey, S.: Principles of correlation and regression analysis. J. Pract. Cardiovasc. Sci. **6**(1), 7–11 (2020). https://doi.org/10.4103/jpcs.jpcs_2_20

17. Electricity producers from alternative sources. State Company "Energorynok" Homepage (2019). http://www.er.gov.ua/. Accessed 03 Nov 2020

18. Ukrainian Assossiation of Renewable Energy Homepage. https://uare.com.ua/en/. Accessed 06 Nov 2020

19. The Global Innovation Index (GII) Homepage. https://www.globalinnovationindex.org/Home. Accessed 03 Nov 2020

20. Wright, T.P.: Factors affecting the cost of airplanes'. J. Aeronaut. Sci. **3**, 122–128 (1936). https://arc.aiaa.org/doi/10.2514/8.155. Accessed 07 Nov 2020

21. Memorandum "Of Understanding on a Strategic Energy Partnership between Ukraine and the European Union in association with the European Atomic Energy Community". The Verkhovna Rada of Ukraine. http://zakon3.rada.gov.ua/laws/show/984_003-16. Accessed 10 Nov 2020

22. ISO and energy. ISO. https://www.iso.org/files/live/sites/isoorg/files/store/en/PUB100320.pdf. Accessed 5 Nov 2020

23. National Power Company Ukrenergo Homepage. https://ua.energy/en/. Accessed 15 Nov 2020

24. The Ministry of Energy will cooperate with the Korean corporation CT in the field of "smart" electricity metering. The Ministry of Energy Homepage. http://mpe.kmu.gov.ua/minugol/control/uk/publish/article?art_id=245485483&cat_id=35109. Accessed 03 Nov 2020

25. DTEK Homepage. https://dtek.com/en/. Accessed 15 Nov 2020

26. Tethys Homepage. https://tethys.pnnl.gov. Accessed 12 Nov 2020

27. Whiting, J., Copping, A., Freeman, M., Woodbury, A.: Tethys knowledge management system: working to advance the marine renewable energy industry. Int. Mar. Energy J. **2**, 29–38 (2019). https://doi.org/10.36688/imej.2.29-38

28. energy trends. ECO-ENERGY Company Homepage. https://ecoenerhiia.ua/news/7-trendiv-energetiki.html. Accessed 12 Nov 2020

29. Western Ukrainian Bioenergy Market Study. PPV Knowledge Networks. https://www.ppv.net.ua/uploads/work_attachments/Western_Ukrainian_Bioenergy_Market_Study_2017.pdf. Accessed 11 Nov 2020

30. Rentechno Homepage. https://rentechno.ua/en/. Accessed 02 Nov 2020

31. WindFarm Homepage. https://www.windfarm.com.ua/. Accessed 02 Nov 2020

32. Institute of Renewable Energy of the National Academy of Sciences of Ukraine Homepage. https://www.ive.org.ua/. Accessed 11 Nov 2020

33. Ram, M., et al.: Global Energy System based on 100% Renewable Energy–Power Sector. Study by Lappeenranta University of Technology and Energy Watch Group, Lappeenranta, Berlin (2017)
34. The Ukrainian Chamber of Commerce and Industry Homepage. https://ucci.org.ua/en/. Accessed 11 Nov 2020
35. European energy management training program starts in Ukraine. Check Point Business Media. https://ckp.in.ua/. Accessed 08 Nov 2020
36. Trofymenko, O.: Transformation effects of the energy sphere in providing economic stability of Ukraine. Econ. Bull. NTUU "KPI" **16**, 49–57 (2019)
37. Tashcheiev, Y.: Prediction of the cost of solar power plants based on the s-curve of technology development and Wright's mathematical model of cost. In: Materials of the XIX International Scientific and Practical Conference, pp. 26–28. Institute of Renewable Energy NAS of Ukraine, Ukraine, Kyiv (2018)
38. Ukrainian Wind Energy Assossiation Homepage. http://uwea.com.ua/en/. Accessed 01 Nov 2020
39. Koba, N., Koba, M.: Forms of cooperation on business. Entrep. Innov. **7**, 75–79 (2019). https://doi.org/10.37320/2415-3583/7.12
40. Renewable Energy Statistics 2020. The International Renewable Energy Agency (IRENA) Homepage. https://www.irena.org/publications/2020/Jul/Renewable-energy-statistics-2020. Accessed 15 Nov 2020
41. Porter, E.: Competitive Strategy: Techniques for Analyzing Industries and Competitors Illustrated Edition, Kindle Edition (2008)
42. Zdrok, V., Lahotskyi, T.: Econometrics. Znannia, Kyiv (2010)

Security in Smart Cities

A Research Perspective on Security in Fog Computing Through Blockchain Technology

Disha Garg$^{(\boxtimes)}$ ⓘ, Komal Kumar Bhatia$^{(\boxtimes)}$ ⓘ, and Sonali Gupta$^{(\boxtimes)}$ ⓘ

J.C. Bose Institute of Science and Technology, YMCA, Faridabad, Haryana, India

Abstract. Most modernistic applications comprising of the Internet of Things are deployed on a cloud environment, which offers advantages like virtual infinite resources, massive elasticity of resources on a pay as per usage metric model. Additionally, IoT real-time applications, essentially self-driving cars, smart home security systems etc., require low latency and less response time. To achieve that, the computation and analysis of data needs to be deployed closer to the edge of the network. The extension of computing services to the edge nodes is termed as 'Fog Computing'. Since every technology comes with two-faced implications; fog computing also suffers with few security issues, majorly comprising user authentication, access control, and data protection from security attacks. Blockchain technology is a promising paradigm to address these security flaws. In this paper, an infrastructure based on blockchain technology coupled with fog computing is proposed as a measure to mitigate the security attacks.

Keywords: Cloud computing · Edge computing · Fog computing · Internet of Things (IoT) · Blockchain technology · Authentication

1 Introduction

The Cisco annual report (2018–2023) [1] states that Internet of Things (IoT) state-of-the-art applications for instance, connected home system will constitute the largest share (approx. 48%) in machine to machine connections; smart car driving will be the rapidly growing application type by the year 2023. Similarly, applications such as healthcare monitoring, asset tracking, video surveillance etc., incorporating billions of connected devices will give rise to the abundance of data, which needs to be managed and analyzed effectively. In contemplating addressing the issue of handling and analyzing big data, Cloud Computing proved to be an up-and-coming solution. On the whole, it uses three types of deployment models; Infrastructure as a Service (IaaS), Platform as a Service (PaaS) and Software as a Service (SaaS); supporting the end-users with virtually unlimited compute resources on a chargeable basis [2]. The cloud computing paradigm uses centralized cloud server clusters, which process the big data and subsequently provide the end users' utilities as per the requirements.

© Springer Nature Switzerland AG 2021
A. Solanki et al. (Eds.): AIS2C2 2021, CCIS 1434, pp. 91–104, 2021.
https://doi.org/10.1007/978-3-030-82322-1_7

Although cloud computing provides data analytics techniques to manage and analyze the massive amount of data, applications requiring real-time data processing do not benefit from the cloud. Since data is stored at a centralized server and is physically located far away from the end-users, it frequently leads to increased transmission latency and increased response time for the time critical applications to reach the end-user [2, 3]. Such real-time applications essentially need the data to be processed closer to the edge nodes. This type of processing is given the term 'Fog Computing'. As defined by Cisco [4], fog computing is an extension of cloud computing, which brings the cloud services closer to the 'things' that generate and analyze the IoT data. A fog node (routers, gateways, switches, embedded servers) can be placed anywhere in a network comprising compute, network, and storage capabilities. Fog computing places the cloud services at a one-hop distance from the user [5] (see Fig. 1).

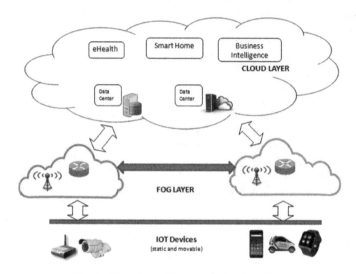

Fig. 1. Overview of fog computing paradigm

Mission-critical applications such as control of computer regulated aircraft flight, cloud robotics, antilock braking system in smart cars etc., requires real-time processing of data closer to the edge to avoid network latency. Such applications exploit the fog computing paradigm to provide a faster response to the actuators [6]. Fog computing offers benefits such as real-time service support, reduced overall network congestion, processing of data from heterogeneous devices, cost-effectiveness, lower power consumption, mobility support [6]. In spite of the said advantages, fog computing suffers from various kinds of security threats. Since different and heterogeneous devices are coupled to the fog nodes in a network; each device is prone to security attack. The hacker can fake the device's IP address to gain unauthorized access to the end user's data [7]. Additionally, each fog node is a potential site of security attack; hence intercommunication among fog nodes needs to be particularly secured. Although security protocols and measures are present in the cloud environment, the same security techniques cannot be applied at the fog level due to the fundamental differences between the two paradigms [7, 8].

Based on the available literature, major security issues in fog computing constitute user authentication, access control, intrusion detection, and data privacy.

To cope with the above protection challenges, an immutable, shared and a distributed community is needed that may efficiently cope with the information transactions. An infrastructure resting on the principles of blockchain is supplied in an attempt to cope with those facts protection-associated troubles. The blockchain network ensures the immutable, tamper-evidence, obvious, and relaxed transactions with the aid of making use of hash values and consensus agreement [9]. With the advent of Blockchain Distributed Network in the field of Information Technology, several sectors inclusive of finance, IoT, governance, education etc. have been benefitted. At present, it is also being endorsed within the healthcare enterprises [10]. The blockchain offers a dispensed and decentralized system that advantages the users and the service carriers. Thus, the blockchain has the capability to reduce the storage fee by as much as 80%. Another utilitarian of using blockchain is diminishing the storage cost of the tremendous amount of IoT facts. [11] analyzed that the storage cost of Amazon S3 is $25 according to TB/month, in contrast to the Blockchain, wherein the cost is about $2 in keeping with TB/month.

The remaining paper is arranged as follows: upcoming segment provides a quick scan into the precedent work submitted in the discipline of fog computing and the various security measures adopted to avoid secure data transmission over fog environment Blockchain Technology. An overview of the security issues and the techniques adopted so far to mitigate user authentication issues, access control, and data protection, faced in fog computing, is described in the next section. The following section describes Blockchain technology with its advantages. Next portion of the paper explains the blockchain's architecture supported fog computing environment for operating the data transactions in a shielded approach. The subsequent module concludes the discussion.

2 Related Works

The potential of fog enabled computing to provide low latency for real-time applications discussed in the previous research articles. Sunyaev et al. [2] explained the key differences in cloud and fog computing. The author discussed the distinction in the said techniques on the parameters such as architecture (centralized and distributed), latency (high and low), support for mobility (limited and supported) and security (low and high), respectively. Puthal et al. [7] depicted the fog computing in three tier architecture, which comprises IoT devices, middleware (transmission medium), and fog servers. [7] also discussed the various security attacks and the corresponding solutions for each tier of the architecture. A research perspective on fog computing has been presented by Bermbach et al. [8]. The authors classified the obstacles faced in this paradigm into two categories. Inherent obstacles include lack of standardized software, increased management effort for managing billions of fog nodes, maintaining the quality of service and nil network transparency. Rest is external obstacles, which includes physical security threats, legal requirements for sensitive information such as healthcare data. Mahmud et al. [14] critically analyze fog-based computing with respect to cloud computing and explains the advantages of the said technique compared to the earlier cloud computing technique.

The researchers also identified specific challenges faced in state of the art applications such as authentication issues, denial of service attacks, privacy disclosure. Alrawais et al. [15] discussed the security challenges faced in fog computing's integrated environment with IoT devices. The authors mainly focused on the authentication issues faced in the fog environment. Yi et al. [16] also explains the various security challenges inherent in fog computing, such as handling an enormous volume of heterogeneous data generating from different smart devices, the requirement of mobility support, authentication issue, rogue fog node, privacy issues and intrusion detection. Romana et al. [17] presents a comprehensive analysis of the various security challenges faced in mobile edge computing. They divided their research in two different parts. The first part identifies the security issues in the edge-based computing architecture. The issues include lack of authentication and identity validity techniques, access control systems, network security, intrusion detection and privacy concerns. Subsequently, the researchers presented possible solutions for each identified issue.

It is evident from the available literature that fog computing suffers from various kinds of security attacks. One possible and effective technique to counter these attacks is through the employment of blockchain technology. Liu et al. [18] exploited the blockchain technology to address the issue of data leakage of IoT data. They proposed a model that uses blended technologies of blockchain, attribute attribute-based access control, and least significant bit for encryption, in an attempt to prevent data privacy attacks. Islam et al. [19] put forward a blockchain-based activity recognition scheme for a smart healthcare system. The security to different and numerous heterogeneous health monitoring devices is provided through blockchain technology. [19] claimed that blockchain technology could control and verify the smart devices, which are responsible for data generation. Jang et al. [20] also exploits the security benefits of blockchain in fog enabled computing for Industrial applications. The authors claimed that blockchain prevents data forgery. Similarly, smart city infrastructure is built using the concepts of fog-enabled computing. The data from various smart devices such as surveillance cameras, traffic lights, alarm systems etc., are transferred to the fog nodes via blockchain. The data travels in the network in the form of data blocks containing encrypted data and hence, protected from unauthorized access.

Jiang et al. [10] worked with non-public Electronic Health Records in diverse arrangements and from diverse assets. The framework was proven on the chained methodology of blockchain to offer privateness and legitimacy of the statistics efficaciously. Griggs et al. [12] innovated the non-public blockchain to permit the use of sensors competently and securely and removed the dangers affiliated to the far off affected person control systems. The authors supplied the patient tracking in actual time to trace the victims from remote locations and conserved the archives of all of the occasions and transactions. A blockchain-based module named 'BlocHIE' proposed a healthcare system for monitoring most cancers sufferers is presented through Shubbar [13]. It adopts contracts for managing the safety of the facts on the healthcare centers and at their residences.

3   Security Issues in Fog Computing

Although fog computing offers various benefits for time-critical applications, it is still prone to security attacks. According to [3], major security attacks in fog computing as shown in Fig 2, constitute Authentication, Privacy, Data Protection, Trust, Access Control and Intrusion Detection. The primary security attack occurs due to lack of proper authentication of user; limited/no access control and no data protection, which are discussed further.

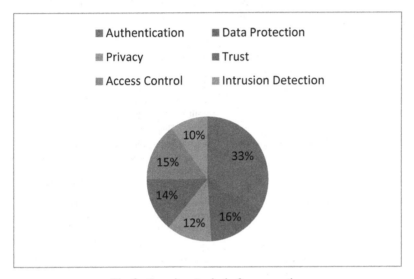

Fig. 2.   Security attacks in fog computing

3.1   Authentication

The distributed characteristic of the fog computing paradigm makes data security as the prime concern. The -users access the smart devices (mobile phones, laptops, PCs, etc.) to retrieve real–time data after being processed and analyzed by the fog nodes. Since billions of diversified devices are coupled in a network, proper authentication of end-users accessing these devices becomes an essential requirement in fog architecture. Fog computing is prone to numerous security attacks, essentially spoofing, physical theft of the smart device, man in the middle attack, impersonation, brute force in password guessing, etc. In an attempt to limit such attacks, the following techniques, as mentioned in Table 1 have been proposed by the researchers.

Table 1. Authentication techniques in fog computing

Reference	Technique used	Advantages	Limitations
Chen et al. 2020 [21]	Authenticated and key exchange scheme	Effective against powerful attacks such as Ephemeral Secret leakage attack	The effectiveness of the proposed technique is proved against the limited attacks, hence not a diversified technique
Jia et al. 2019 [22]	Mutual authentication supporting anonymity and un-traceability in mobile edge computing	Identity-based authorization requiring only one round of message exchange to validate authentication	The technique is not robust or adaptive in case of regulatory changes
Zhu et al. 2019 [23]	Authentication for smart Grid application in fog environment	The technique offers an effective privacy-preserving mechanism for user's sensitive data. It also provides data aggregation	The proposed technique is not effective in preventing collision attacks
Wasid et al. 2018 [24]	Secure key agreement scheme in fog computing	Preserves anonymity and un-traceability properties for protecting user's sensitive data	The scheme is not effective in diminishing the storage overheads over the network
Jia et al. 2018 [25]	Three party Authenticated Key Agreement protocol	Protocol maintains user anonymity and un-traceability	Not useful in case of Ephemeral Secret leakage attack
Imine et al. 2018 [26]	MASFOG: Mutual Authentication scheme for fog computing, which is based on secret sharing and blockchain technology	Fog nodes and users mutually authenticate each other without resorting to the cloud. Also, fog nodes show one another at the network edge using blockchain Most Effective in handling impersonating attacks	Do not address the revocation problem in the fog environment
Shen et al. 2018 [27]	Matrix-based key agreement scheme in the healthcare system	Provides lightweight authentication between patient and doctors and also between patients and fog sensors	The proposed technique provides security against limited attacks such as man in the middle attack, replay attack, un-traceability and anti-counterfeiting

3.2 Access Control and Data Protection in Fog Computing

Applications requiring real-data data such as smart healthcare, vehicular systems, smart home etc., leverage the benefits of Fog computing paradigm, where the end user's private and sensitive data is transferred to the fog nodes from the edge devices (mobile phone, smart watch, wearable health monitors, self driving car, etc.). Since the edge devices are situated at the end of the network, it elevates serious security apprehensions, which can be dealt with stringent access controls and data protection. For instance, in a health care system, proper access privileges must be set up for patients, doctors and other stakeholders to ensure that patient's medical records are safe and protected from any unauthorized access. Table 2 lists the various techniques that are exploited in an attempt to provide adequate access control and data protection techniques.

Table 2. Access control and data protection techniques in fog computing

Reference	Technique	Advantages	Limitations
Altulyan et al. 2020 [28]	Securing integrity of data in smart cities using blockchain technology and secret sharing of sensitive information	Data is transferred from one to another node in a network which is hidden from the nodes. The hiding of data is achieved by secret sharing and blockchain technology	The technique does not address the issues of network latency and network bandwidth
Fan et al. 2019 [29]	Privacy-preserving multi-authority access control using attribute-based encryption	One way key agreement keeping the attributes anonymous, to preserve the privacy of users Computational overhead is reduced on the user side since the decryption procedure is outsourced	Security of data needs to be taken into account while protecting privacy
Xue et al. 2018 [30]	An efficient and fine-grained access control in a fog based vehicular cloud computing environment	The technique effectively pushes the vehicle's mobility information to the fog server from the cloud server to reduce latency and provide protection to confidential data while outsourcing	The technique covers a limited domain of security attacks
Piao et al. 2018 [31]	Differential privacy approach for preserving privacy of user's sensitive data in a government cloud	The data privacy-preserving algorithm reduces query sensitivity from and to the government cloud	The proposed algorithm was not tested against experimental/ real data sets
Wang et al. 2018 [32]	Three-layer privacy-preserving approach in fog architecture	Storage of data is achieved by partitioning data into different cloud, fog, and local storage servers, hence achieving data privacy	Lack of experimental data to support the effectiveness of the technique against privacy attacks
Yu et al. 2017 [33]	Secure and leakage-resilient, functional encryption scheme	Security against side-channel attacks	The technique is not effective against attacks other than side-channel attacks

4 Blockchain Technology

The Blockchain technology is built on a distributed framework with p2p connectivity in a dispensed environment and it stepped in the year 2008. The initial usage of the technology was the crypto currency referred to as Bitcoin [34]. Subsequently, it is been used in several sectors. Everyday scholars are investigating about the unexplored possibilities and domain names of blockchain programs. A Blockchain operates as a log that assists the dispensed network for conserving the information in a distributed manner. The transactional information is labeled and saved like a series in blocks linked using hash values. Haber [35] first set up the Blockchain implementation and explained how the blocks have been connected and created in a secure fashion. The same methodology was upgraded by Bayer et al. [36] through introducing the notion of Merkel Tree for enhancing system proficiencies and storage expandability. Although Blockchain is dominantly used within the economic quarter, its capability to make sure immutable and apparent transactions and maintain ancient information has given it the threat to outrank and be accompanied with the aid of multiple extraordinary domain names inside the blockchain community, which need no middleman and are automatic. The transactions are achieved through the usage of the idea of astute contacts, which attempts to solve some mathematical computations, only after which the transaction is said to have been progressed. Furthermore, the consensus of the network is likewise preserved. For instance, when a transaction is asked, then it will only exceed if all of the network nodes comply with it [37].

A block sequence which is related collectively is referred as a Blockchain. The structure of the block consists of the Header and Body (see Fig. 3). The header includes details similar to the blockchain's version, previous block's header of the chain, timestamp, Merkel tree, metadata, and plenty of others. The set of rules pertaining to a data block is defined by the version. Merkle tree is customary for each transaction to ensure that the facts are not mutable [36]. Additionally, virtual signatures or hashing enables security of the transactions. Furthermore, a block is related to the preceding block by the use of the hashing value. Therefore, a block incorporates its very own hash price besides the hashing charge of the preceding block. Hence, a blockchain is consistent. The initial block of the chain is termed as the header block. During an event of a transaction request, a new block is created and statistics resides for a particular duration in that block's memory [38, 39]. Afterwards, Merkle tree is created to preserve the transactional element within the frame of the block. The value of the hashing function is encapsulated in the header of the block using some cryptographic strategies. Hence, each single block contains its separate and unique hash value, which helps in maintaining a sequenced chain. Post the production of the hashing values, the timestamp of the transactions are generated and consequently, the block is ultimately formed [39].

The Blockchain infrastructure is cut up into different levels, as depicted in Fig. 4 underneath. Each level or the layer has its separate and a dedicated function. The foundation layer constitutes the data layer, accountable for collecting the data from varied gadgets (hardware) and enclosing it for protection. Hashing, virtual signatures, Merkle tree, cryptographic technologies [37] are applied on this layer. Network layer is the subsequent layer accountable for sustaining the p2p connections between distributed nodes of the network. Next comes the Consensus layer, which offers the consensus agreements

BLOCKCHAIN

DATA   📄 Welcome to Blockchain Demo 2.0!

PREVIOUS HASH 0

HASH   000de75a315e77a1f9e98fb6247d03dd18ac52632d7dc6e9920261d8109b37cf

GENESIS BLOCK on Tue, 17 Oct 2017 19:53:20 GMT   604

⌄

DATA   📄

PREVIOUS HASH 000de75a315e77a1f9e98fb6247d03dd18ac52632d7dc6e9920261d8109b37cf

HASH   000be3e60e73b4cf78048b952fd726179c7581d6afcd2aa70479c32e80261ea

BLOCK #1 on Sat, 10 Feb 2018 15:40:00 GMT   3244

⌄

Fig. 3. Snapshot of a blockchain

in the device; and hence assuring the trustworthiness of the nodes [40]. The above mentioned three layers are said to be the necessary layers within a blockchain structure. In assessment, a non-obligatory layer can be present incorporating the required functionalities, if needed. The infrastructure also incorporates smart contracts for enabling well ordered transactions with none middlemen.

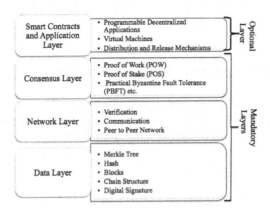

Fig. 4. Blockchain layered architecture

4.1   Advantages of Blockchain Technology

Increased Transparency
Blockchain is a distributed ledger therefore, all the nodes in a network share a same data documentation instead of individual and separate copies. The consensus protocol in a blockchain is responsible for updating the shared copy of the information, i.e., once every node in the network comes to a consensus. Only a new block of data can be added in the chain of blocks. To alter or change a single transaction, the entire blockchain needs to be changed in the network. Hence, data residing in a blockchain is more transparent and consistent than other record-keeping systems. The same data is visible to all the legitimate nodes of the network.

Greater Security
Each new block of data is added to the blockchain using the previously added block's hash value. Additionally, the transactions must be agreed upon by the entire nodes in a network before they are recorded. The information is stored across a network of nodes (computers) instead of a single centralized server. Hence it is strenuous for the adversary to compromise the data. Organizations dealing with sensitive information such as healthcare, government agencies, financial services greatly benefit from blockchain technology.

Peer–to-Peer Network
Information is shared and recorded in all the nodes in the network. Hence the network becomes more robust with the increasing number of nodes. Each node is considered equal and equally important for keeping the information safe and secure.

Trust
The identity of the participating nodes in a blockchain based network remains confidential. Hence every node is free to communicate with the rest of the nodes via the secure network. This leads to anonymity and security of user's data, which are the most useful advantages of blockchain technology [41].

5   Blockchain in Fog Computing Environment

The advantages mentioned above, makes blockchain technology a promising paradigm for ensuring the security of sensitive information of users travelling in a fog environment. As depicted in Fig. 5, every fog node is a potential node for the Blockchain, which can securely distribute and preserve the users' authentication and authorization information by employing the secret keys. Blockchain features such as immutability and time-stamping ensure transparency in the transactions of data.

The raw data is generated and collected from multiple (static and movable) IoT devices, for instance, mobile phones, surveillance cameras, health detectors, wearable devices such as BP monitor, pulse monitor, etc., in various heterogeneous formats. After the required pre-processing and labeling of the raw data, it is transferred to the fog node for storage and further strategic analysis. In order to assure security and integrity of the user's sensitive information such as health records, banking information, etc., the stored data and the users' personal details are made encrypted exploiting certain cryptographic strategies and virtual signatures. Along with the cryptographic techniques (public and private key), proper authentication techniques are exploited to guarantee that only permissible users can access and send the data to the fog network.

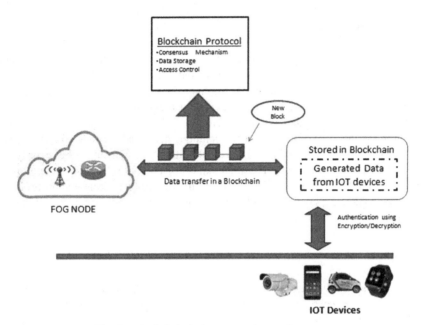

Fig. 5. Blockchain in fog computing environment

The encrypted data travels in the network and reach the fog nodes in the form of blocks. *Miners* are present in the fog environment and are accountable for controlling the security aspect of the network by catering for the responsibility of keeping the secret shared keys guarded. Once a transaction is initiated, it is responsibility of the miners for validating the transaction. The transactions cannot enhance without their compliance. Further, miners are also responsible for manifesting the transactions by acting complicated computations, hence making sure consensus mechanisms are followed.

The prime point of the said exercise is to set forth a framework for dealing with the records throughout the data transaction activities and also during residing stage of the data at the fog nodes. Hence, the proposed framework is useful for protecting and securing the information in transit as well as at rest. Only authorized users can be granted access while the data is presiding over the fog node. Moreover, the data is saved in concealed shape, therefore in the event of an information breach; statistics' security might no longer be compromised. Furthermore, the key might be shared to the blockchain's nodes in a dispensed approach that could diminish or eliminate a sole supervising authority, failing which would cause information privacy-associated troubles. Data residing in the transaction is secured through the blockchain by means of administering the hashing value in addition to exploiting the consensus agreements. Furthermore, customers' identity is kept anonymous along with preserving the integrity of the user's sensitive data because the blockchain helps in time embossing the data. Therefore, adopting this sort of a framework might assure the safety and privacy of the personal records in a fog environment and provide actual-time information get admission to helping in faster reaction time for time crucial programs.

6 Conclusion

In this paper, a systematic and comprehensive review of fog computing and various security issues in fog computing has been presented. Major security loop holes originate from lack of proper user authentication techniques, limited access control, and sensitive data loss. To mitigate the said issues, blockchain based security technique has been presented that protects the data on the fog environment from unauthorized access, also maintaining the integrity of the data. This technique gives end users the ability to share the data on the network consisting of billions of nodes while enhancing the security and seclusion of user's sensitive data. The amalgam of fog computing and the technology of blockchain is capable of coping with the scalability difficulty, making it appropriate for time-crucial applications which includes Healthcare, Banking, Self-using cars, Robotics, and many others big volumes of heterogeneous data.

References

1. Cisco Annual Internet Report (2018–2023) White Paper. https://www.cisco.com/c/en/us/solutions/collateral/executive-perspectives/annual-internet-report/white-paper-c11-741490.html. Accessed Sep 2020
2. Sunyaev, A.: Fog and egde computing. In: Internet Computing. Springer, Cham (2020). https://doi.org/10.1007/978-3-030-34957-8_7
3. Kaur, J., Agrawal, A., Khan, R.A.: Security issues in fog environment: a systematic literature review. Int. J. Wirel. Inf. Netw. 27(3), 467–483 (2020). https://doi.org/10.1007/s10776-020-00491-7
4. Fog Computing and the Internet of Things: Extend the Cloud to Where the Things Are. https://www.cisco.com/c/dam/en_us/solutions/trends/iot/docs/computing-overview.pdf. Accessed Sep 2020
5. Kunal, S., Saha, A., Amin, R.: An overview of cloud-fog computing: architectures, applications with security challenges. Secur. Priv. 2, e72 (2019). https://doi.org/10.1002/spy2.72
6. Gupta, H., Vahid Dastjerdi, A., Ghosh, S.K., Buyya, R.: iFogSim: a toolkit for modeling and simulation of resource management techniques in the Internet of Things, edge and fog computing environments. J. Softw.: Pract. Exp. 47, 1275–1296 (2017). https://doi.org/10.1002/spe.2509
7. Puthal, D., Mohanty, S.P., Bhavake, S.A., Morgan, G., Ranjan, R.: Fog computing security challenges and future directions [Energy and Security]. IEEE Consum. Electron. Mag. 8(3), 92–96 (2019). https://doi.org/10.1109/MCE.2019.2893674
8. Bermbach D., et al.: A research perspective on fog computing. In: Braubach, L., et al. (eds.) Service-Oriented Computing – ICSOC 2017 Workshops. ICSOC 2017. Lecture Notes in Computer Science, vol. 10797. Springer, Cham (2017). https://doi.org/10.1007/978-3-319-91764-1_16

9. Coutinho, E.F., Paulo, D.E., Abreu, A.W., Carla, I.M.B.: Towards cloud computing and blockchain integrated applications. In: 2020 IEEE International Conference on Software Architecture Companion (ICSA-C), Salvador, Brazil, pp. 139–142 (2020). https://doi.org/10.1109/ICSA-C50368.2020.00033

10. Jiang, S., Cao, J., Wu, H., Yang, Y., Ma, M., He, J.: BlocHie: a blockchain-based platform for healthcare information exchange. In: Proceedings of the 2018 IEEE International Conference on Smart Computing (SMARTCOMP), Taormina, Italy, 18–20, June 2018 (2018)

11. Singh, P., Nayyar, A., Kaur, A., Ghosh, U.: Blockchain and fog based architecture for Internet of everything in smart cities. Future Internet 12(4), 61 (2020)

12. Griggs, K.N., Ossipova, O., Kohlios, C.P., Baccarini, A.N., Howson, E.A., Hayajneh, T.: Healthcare blockchain system using smart contracts for secure automated remote patient monitoring. J. Med. Syst. 42, 130 (2018)

13. Shubbar, S.: Ultrasound medical imaging systems using telemedicine and blockchain for remote monitoring of responses to neo adjuvant chemotherapy in women's breast cancer: concept and implementation. Master's Thesis, Kent State University, Kent, OH, USA (2017)

14. Mahmud, R., Kotagiri, R., Buyya, R.: Fog computing: a taxonomy, survey and future directions. In: Di Martino, B., Li, K.-C., Yang, L.T., Esposito, A. (eds.) Internet of Everything. IT, pp. 103–130. Springer, Singapore (2018). https://doi.org/10.1007/978-981-10-5861-5_5

15. Alrawais, A., Alhothaily, A., Hu, C., Cheng, X.: Fog computing for the Internet of Things: security and privacy issues. IEEE Internet Comput. 21(2), 34–42 (2017). https://doi.org/10.1109/MIC.2017.37

16. Yi, S., Qin, Z., Li, Q.: Security and privacy issues of fog computing: a survey. In: Xu, K., Zhu, H. (eds.) WASA 2015. LNCS, vol. 9204, pp. 685–695. Springer, Cham (2015). https://doi.org/10.1007/978-3-319-21837-3_67

17. Roman, R., Lopez, J., Mambo, M.: Mobile edge computing, Fog et al.: a survey and analysis of security threats and challenges. Future Gener. Comput. Syst. 78, 680–698 (2018). https://doi.org/10.1016/j.future.2016.11.009

18. Liu, Y., Zhang, J., Zhan, J.: Privacy protection for fog computing and the Internet of Things data based on blockchain. Clust. Comput. 24(2), 1331–1345 (2020). https://doi.org/10.1007/s10586-020-03190-3

19. Islam, N., Faheem, Y., Din, I.U., Talha, M., Guizani, M., Khalil, M.: A blockchain-based fog computing framework for activity recognition as an application to e-Healthcare services. Future Gener. Comput. Syst. 100, 569–578 (2019). https://doi.org/10.1016/j.future.2019.05.059

20. Jang, S.-H., Guejong, J., Jeong, J., Sangmin, B.: Fog computing architecture based blockchain for industrial IoT. In: Rodrigues, J.M.F., et al. (eds.) ICCS 2019. LNCS, vol. 11538, pp. 593–606. Springer, Cham (2019). https://doi.org/10.1007/978-3-030-22744-9_46

21. Chen, C.-M., Huang, Y., Wang, K.-H., Kumari, S., Wu, M.-E.: A secure authenticated and key exchange scheme for fog computing. Enterp. Inf. Syst. (2020). https://doi.org/10.1080/17517575.2020.1712746

22. He, D., Choo, K.-K.R., Jia, X., Kumar, N.: A provably secure and efficient identity-based anonymous authentication scheme for mobile edge computing. IEEE Syst. J. 14, 560–571 (2019). https://doi.org/10.1109/JSYST.2019.2896064

23. Zhu, L., et al.: Privacy-preserving authentication and data aggregation for fog-based smart grid. IEEE Commun. Mag. 57(6), 80–85 (2019). https://doi.org/10.1109/MCOM.2019.1700859

24. Wazid, M., Das, A.K., Kumar, N., Vasilakos, A.V.: Design of secure key management and user authentication scheme for fog computing services. Future Gener. Comput. Syst. 91, 475–492 (2019). https://doi.org/10.1016/j.future.2018.09.017

25. Jia, X., He, D., Kumar, N., Choo, K.-K.: Authenticated key agreement scheme for fog-driven IoT healthcare system. Wirel. Netw. **25**(8), 4737–4750 (2018). https://doi.org/10.1007/s11 276-018-1759-3
26. Imine, Y., Kouicem, D.E., Bouabdallah, A., Ahmed, L.: MASFOG: an efficient mutual authentication scheme for fog computing architecture. In: 17th IEEE International Conference on Trust, Security and Privacy in Computing and Communications/12th IEEE International Conference on Big Data Science and Engineering (TrustCom/BigDataSE), New York, NY, pp. 608–613 (2018). https://doi.org/10.1109/TrustCom/BigDataSE.2018.00091
27. Shen, J., Yang, H., Wang, A., Zhou, T., Wang, C.: Lightweight authentication and matrix-based key agreement scheme for healthcare in fog computing. Peer-to-Peer Netw. Appl. **12**(4), 924–933 (2018). https://doi.org/10.1007/s12083-018-0696-3
28. Altulyan, M., Yao, L., Kanhere, S.S., et al.: A unified framework for data integrity protection in people-centric smart cities. Multimed. Tools Appl. **79**, 4989–5002 (2020). https://doi.org/10.1007/s11042-019-7182-2
29. Fan, K., Xu, H., Gao, L., Li, H., Yang, Y.: Efficient and privacy preserving access control scheme for fog-enabled IoT. Future Gener. Comput. Syst. **99**, 134–142 (2019). https://doi.org/10.1016/j.future.2019.04.003
30. Xue, K., Hong, J., Ma, Y., Wei, D.S.L., Hong, P., Yu, N.: Fog-aided verifiable privacy preserving access control for latency-sensitive data sharing in vehicular cloud computing. IEEE Netw. **32**(3), 7–13 (2018). https://doi.org/10.1109/MNET.2018.1700341
31. Piao, C., Shi, Y., Yan, J., Shang, C., Liu, L.: Privacy-preserving governmental data publishing: a fog-computing-based differential privacy approach. Future Gener. Comput. Syst. **90**, 158–174 (2018). https://doi.org/10.1016/j.future.2018.07.038
32. Wang, T., Zhou, J., Chen, X., Wang, G., Liu, A., Liu, Y.: A three-layer privacy preserving cloud storage scheme based on computational intelligence in fog computing. IEEE Trans. Emerg. Top. Comput. Intell. **2**(1), 3–12 (2018). https://doi.org/10.1109/TETCI.2017.2764109
33. Yu, Z., Au, M., Xu, Q., Yang, R., Han, J.: Towards leakage-resilient fine-grained access control in fog computing. Future Gener. Comput. Syst. **78**, 763–777 (2018). https://doi.org/10.1016/j.future.2017.01.025
34. Crosby, M., Pattanayak, P., Verma, S., Kalyanaraman, V.: Blockchain technology: beyond bitcoin. Appl. Innov. **2**, 71 (2016)
35. Haber, S.A., Jr. Stornetta, W.S.: Method for Secure Time-Stamping of Digital Documents. US5136647A (1992)
36. Bayer, D., Haber, S., Stornetta, W.S.: Improving the efficiency and reliability of digital time-stamping. In: Capocelli, R., De Santis, A., Vaccaro, U. (eds.) Sequences II, Springer, New York, pp. 329–334 (1993). https://doi.org/10.1007/978-1-4613-9323-8_24
37. Xia, Q., Sifah, E.B., Asamoah, K.O., Gao, J., Du, X.: MeDShare: trust-less medical data sharing among cloud service providers via blockchain. IEEE Access **5**, 14757–14767 (2017). https://doi.org/10.1109/ACCESS.2017.2730843
38. Sharma, P.K., Moon, S.Y., Park, J.H.: Block-VN: a distributed blockchain based vehicular network architecture in smart City. J. Info. Proc. Syst. **13**(1), 84 (2017)
39. Jiao, Y., Wang, P., Niyato, D., Suankaewmanee, K.: Auction mechanisms in cloud/fog computing resource allocation for public blockchain networks. IEEE Trans. Parallel Distrib. Syst. **30**(9), 1975–1989 (2019). https://doi.org/10.1109/TPDS.2019.2900238
40. Xueping, L., Shetty, S., Tosh, D., Kamhoua, C., Kwiat, K., Njilla, L.: Provchain: a blockchain-based data provenance architecture in cloud environment with enhanced privacy and availability. In: Proceedings of the 17th IEEE/ACM International Symposium on Cluster, Cloud and Grid Computing, pp. 468–477. IEEE Press (2017)
41. Zyskind, G., Nathan, O., Pentland, A.S.: Decentralizing privacy: using blockchain to protect personal data. In: Proceedings of the 2015 IEEE Security and Privacy Workshops (SPW 2015), San Jose, CA, USA, 21–22 May 2015, pp. 180–184 (2015)

ZKPAUTH: An Authentication Scheme Based Zero-Knowledge Proof for Software Defined Network

Hamza Mutaher$^{(\boxtimes)}$ (iD) and Pradeep Kumar (iD)

Department of Computer Science and Information Technology, Maulana Azad National Urdu University, Hyderabad, India

Abstract. To secure the communication between the control and data plane devices and keep the communication channel in software defined network (SDN) immune against network attacks. OpenFlow recommends configuring the transport layer security (TLS). Unfortunately, some OpenFlow devices don't adopt TLS. In this case, SDN needs a robust network authentication protocol to keep the communication within SDN secured. In this paper, we proposed ZKPAUTH, An Authentication Scheme based on Zero-Knowledge Proof to enforce an authentication between the controller and the hosts before establishing the communication. The controller and the hosts take help of the nonce distribution center (NDC) that maintains the authentication between them. We implemented the security of the ZKPAUTH using AVISPA tool. The result of AVISPA shows that the ZKPAUTH is safe against replay and MITM attacks. The security analysis of ZKPAUTH is discussed and proved that ZKPAUTH is efficient and secure against DoS and host impersonation attack.

Keywords: SDN · Controller · Host · OpenFlow · Zero-knowledge proof · Mutual authentication

1 Introduction

Software defined network (SDN) came into view with the idea of segregating the network control logic from the network forwarding device; therefore, it overcomes the network management complexities [1]. Generally, SDN architecture has three layers: control, data, and application planes. The application plane enforces the networking policies into the control plane through the northbound interfaces (APIs) [2]; subsequently, the control plane deploys these networking policies to the data plane devices by the southbound interfaces; OpenFlow protocol [3]. OpenFlow is a standard protocol that connects the SDN control plane to the SDN data plane through the OpenFlow channel. The controller device in SDN is a centralized control plane device with full control and a complete overview of the network.

Hence, the controller is a centralized device; it is considered a single point of failure and might affect network operations [4]. To secure the communication channel between

© Springer Nature Switzerland AG 2021
A. Solanki et al. (Eds.): AIS2C2 2021, CCIS 1434, pp. 105–120, 2021.
https://doi.org/10.1007/978-3-030-82322-1_8

the controller and the data plane device, OpenFlow recommends using TLS. Unfortunately, some tasks of the secured TLS are complicated to be configured on OpenFlow [5, 6]. Moreover, the OpenFlow keeps the TLS configuration as an optional mode of communication which means some OpenFlow-enabled devices do not adopt the TLS [7]. This makes it easy for an attacker to establish a denial of service attack (DoS), man-in-the-middle attack (MITM) host impersonation attack, and other network attacks against the controller.

To overcome the above issues, the ZKPAUTH authentication scheme was proposed to allow both controller and host to authenticate each other before establishing the communication. The proposed scheme uses zero-knowledge proof (ZKP) to implement the authentication process and improve controller security against malicious network users. Goldwasser introduces ZKP, Micali and Rackoff in [8], where Feige, Fait and Shamir in [9] enhanced it to the proof of identity. ZKP is an authentication process that happens among two parties (Alice & Bob). Alice is the prover and Bob is the verifier. The secret among them both is S. The prover has to prove to the verifier that she knows the secret S without exposing the secret itself. The verifier has to verify that the prover knows the secret S or not. ZKP has three characteristics that make it trustworthy to be used in network security:

- Completeness: both the prover and verifier must be honest; the honest verifier must be convinced if the proving statement is true.
- Soundness: no fake prover can convince the verifier if the statement is false.
- Zero-Knowledge: if the statement is true, the verifier does not learn anything other than the statement.

In the proposed scheme, the prover is the network host where the verifier is the network controller.

The sections of this paper are arranged as: in Sect. 2, the related work was discussed. In Sect. 3, the ZKPAUTH with its four phases was explained. In Sect. 4, the implementation of the Zero-Knowledge Proof Authentication (ZKPAUTH) AVISPA tool was demonstrated. Then the security analysis was discussed in Sect. 5. In Sect. 6, the performance of ZKPAUTH was evaluated. Finally, the conclusion was discussed with future work.

2   Related Work

Security of SDN has different aspects; the authentication between control and data planes devices focuses on keeping trusted devices in the network and preventing untrusted ones. Various researches and developments have been conducted to analyze, investigate and enhance the authentication among SDN devices. HiAuth [10] is a high authentication mechanism that has been proposed as an obscure and lightweight authentication mechanism to secure the SDN controller against DoS attack. The mechanism they designed hides the transaction identification (XID) of the forwarding devices in the control packet header and makes it undetectable to the attacker. Wang *et al.* [11] have developed PERM-GUARD, a solution to guarantee SDN flow rules' validity. PERM-GUARD implements a model of authentication permission and introduces a scheme of identity-based signature for allowing the controller to identify the legitimate flow rules from fake ones. The experiment results show that the PERM-GUARD is capable of finding out and denying the malicious flow rules. Moreover, it can filter out the untheorized flow rules created by the legitimate application and identify their creator. Mattos *et al.* [12] introduced Auth-Flow, a technique of authentication and access control in SDN. The technique focuses on the host authentication above the MAC layer of the OpenFlow network. According to the host's privileges level, AuthFlow maps the host credentials with a set of flow rules to implement a credentials-based authentication access control procedure. This technique allows the controller to use the host's identity as a new filed of the flow to identify the forwarding rules. The result shows that AuthFlow can reject the host with fake credentials or unauthorized access rules. Some researches effort to create some new attacks and find countermeasures for those attacks [13]. Hong *et al.* [14] proposed a new unique SDN vectors attack that can poison the SDN controllers and this attack works differently from the legacy network. This attack may lead to serious hijacking, DoS and MITM attacks. The authors suggested that the Floodlight, OpenDaylight, Beacon and POX controllers are vulnerable to network topology poisoning attacks. Then they proposed TopoGuard, a new security extension that secures the controller against the network topology poisoning attack. The new extension can provide real-time detection of the above attack. The experiment of TopoGurad shows that the solution can secure the topology of the network effectively. For inter-domain authentication of SDN Zhou *et al.* [15] presented an authentication protocol for inter-domain SDN controllers. The authentication protocol depends on a third-party trust domain called Trust measurement module (TMM). TMM is designed to measure the controller flow rules and guarantee security in SDN inter-domain scenario. Buragohain *et al.* [16] designed FlowTrApp, a security framework for SDN. This framework was designed to detect the high and low rate of Distributed Denial of Service attack (DDoS) and mitigate it. Mitigation takes action by matching the incoming traffic with the consistent flow traffic rules. Cui *et al.* [17] designed an authentication mechanism for the applications run on the SDN environment. This mechanism applied a system of authentication between the untrusted network and user request. This is considered as one of the key challenges in SDN networks. Cho *et al.* [18] designed a technique for access control based on fine-grained, along with the mutual authentication process. The mutual authentication process uses a cryptographic hash function without the need of installing complicated cryptographic requirements. This mechanism's essential aim is to prevent an authorized packet from accessing the network resources. Lee *et al.* [19] designed a strengthen authentication technique for SDN to verify the network manager's authenticity, which tries to gain access to the controller. This technique's main concept is to forward packets to the controller with a registered path, where the controller accepts packets forwarded through the registered path only. Ellinidou *et al.* [20] designed a new security protocol to secure communication between SDN parties. The new protocol focused on managing the routing procedure over the cloud of chips (CoC). Fang et al. [21] designed and implemented the hidden pattern (THP) to secure the network permissions from being controlled by hackers. This authentication scheme concerns to prevent network permissions from multiple types of authentication attacks using the graphical password and digital challenge at the same time. THP scheme helps the network providers to control their networks from being accessed illegally by unauthenticated users. ZKP is also used as an authentication protocol by Major *et al.* [22] where they used port knocking solutions along with ZKP to protect the severs ports form being scanned by the attackers. Natanzi *et al.* [23] proposed a mechanism that allows multi-controllers to authenticate each other. In this mechanism, the private and public keys are created using elliptic curve cryptography (ECC) for the key management process with the authentication server's help. This mechanism is vulnerable to internal DoS attacks due to less security among data and control planes. In improving the security of the controller in both single control and multiple control domains for the security between the SDN controller and the internet of things (IoT) devices. Ambili *et al.* [24] proposed a secure data sharing system. This system allows the IoT devices to communicate with the SDN controller securely using ciphertext cryptography, and the authorized devices have the correct encryption key to decrypt the ciphertext. Ravindra *et al.* [25] proposed an authentication scheme for SDN. This scheme uses a radius authentication server to help the controller authenticate the network users to access the network resources.

Table 1. ZKPAUTH important notations.

Notation	Description
ID	Host identity
PW	Host password
$h()$	Cryptographic hash function
N	Nonce
R	Random number
Y_j	List of public keys
X_j	List of private keys
C_j	List of challenges
Z_n	Real numbers
NCD	Nonce distribution center

3 ZKPAUTH the Proposed Scheme

The Zero-Knowledge Proof Authentication (ZKPAUTH) main idea is to protect the SDN controller from being accessed by fake identity network hosts and ensure its immunity against network attacks. To ensure that the controller is secured, ZKP was implemented to the SDN controller. In ZKPAUTH, there are three parties of the communication. The host (prover), the controller (verifier) and the nonce distribution center (NDC). The host has to prove the correctness of its identity to the controller. The controller has to verify the host identity correctness. The NDC is responsible for distributing the nonce to establish the host and the controller's identification and verification process.

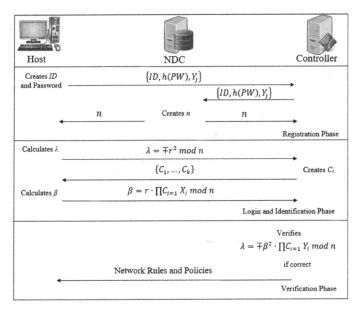

Fig. 1. ZKPAUTH systematic process.

The host has to prove its knowledge about the nonce without revealing the nonce itself, and this is the interesting property of the ZKP. The systematic process of ZKPAUTH is depicted in Fig. 1. ZKPAUTH consists of four phases. The key-generation phase, registration phase, the login and identification phase and the verification phase. In the key generation phase, the host has to create its private and public keys. In the registration phase, the host creates its *ID* and password and send them along with its public keys to the controller as a registration request. Then the controller forwards this request to the NDC and NDC sends a unique nonce to both host and controller.

In the login and identification phase, the host attempts to log in to the SDN network. This attempt is not like the traditional login attempts where the host has to prove its password; instead, it is a ZKP proof where the host proves that it knows the nonce without exposing the real nonce to the controller. Therefore, the host sends a login request to the controller; this request contains a random number and the host's public key. The controller sends back the host a list of challenges to make it prove its identity. The host solves the controller's challenges using its private keys and sending this solution to the controller along with another random number. In the verification phase, the controller verifies the host identity's correctness using random numbers, public keys of the host and the list of the challenges. If the host identity is verified, then the host will be able to get the network rules and policies from the controller and will be able to send and receive network packets as well, if not then the host has to repeat the login attempt. The four phases of the ZKPAUTH are explained in details below, and the important notations of the ZKPAUTH are depicted in Table 1.

Algorithm1: Key-Generation Phase

All communication parties have to agree on $n = (p * q)$.

Public keys Y_j must be $Y_j = 1/X_j^2 (mod\ n)$

Public_Key_Gen ()

	Input:	*Random numbers $X_j = \{X_1, ..., X_k\}$ as private keys*
		$n = (p * q)$ as a nonce
	Output:	*$Y_j = \{Y_1, ..., Y_k\}$ as public keys*
	Procedure:	
	Begin	*{ Host: H*

$$\text{To generate public keys:}$$
$$Y_1 = 1/X_1^2 (mod\ n)$$
$$Y_2 = 1/X_2^2 (mod\ n)$$
$$Y_3 = 1/X_3^2 (mod\ n)$$
$$Y_k = 1/X_k^2 (mod\ n)$$

End *}*

3.1 Key-Generation Phase

In the key-generation phase, the host must create public and private keys to use them later on in the next phases. The host generates its keys as follows:

- The host generates $X_j = \{X_1, \ldots, X_k\}$ randomly where $X_j \in Z_n$
- The host generates $Y_j = \{Y_1, \ldots, Y_k\}$ randomly and independently where $Y_j = 1/X_j^2 (mod\ n)$
- The host publishes $Y_j = \{Y_1, \ldots, Y_k\}$ as public keys and keeps $X_j = \{X_1, \ldots, X_k\}$ as private keys.

3.2 Registration Phase

After generating the public and private keys, the host has to register itself to the controller. Therefore, it will be able to login into the SDN network and get network policies. The registration steps are as follows:

- The host creates the host *ID* and Host password *PW*.
- The host hashes the *PW* using the OWH function $h(.)$ as $h(PW)$.
- The host sends the controller a registration request contains its *D*, hashed password and the public keys as $\{ID, h(PW), Y_j\}$.

Algorithm2: *Registration Phase*
All communication parties have to agree on n $= (p * q)$

Reg_Phase ()

Input:		*ID and* $h(PW)$
		$n = (p * q)$ *as a nonce*
Output:		
Procedure:	*{ Host: H*	
Begin	*Create:*	*ID = Host identity*
		PW = Host password
	Compute:	*PW* $= h(PW)$
	Send:	*C* $\leftarrow$ *(ID,h(PW),Y$_j$)*
	}	
	{ Controller: C	
	Froward:	*NDC* $\leftarrow$ *(ID,h(PW),Y$_j$)*
	}	
	{ NDC:NDC	
	Compute:	$n = (p * q)$
	Send:	*H* $\leftarrow n$
		C $\leftarrow n$
	}	
End		

3.3 Login and Identification Phase

Far from the traditional way of login, i.e., in every login attempt, the host has to provide its password, wherein ZKPAUTH the host has to login into the SDN network using the ZKP proving technique. The host must prove to the controller that it knows the nonce n without exposing the real nonce. The steps of the login and identification phase are as follows:

- The host creates random numbers r and λ calculates $\lambda = \mp r^2 \bmod n$ and sends λ to the controller.
- The controller creates random challenges $C_j = \{C_1, \ldots, C_k\}$ as Boolean vectors and send them to the host.
- The host creates a random number β and calculates $\beta = r \cdot \prod C_{j=1} X_j \bmod n$ and sends the β to the controller.

The third step is the host's identification step to the controller. It is also called the proof of the identity; thus, this step has to be repeated multiple times depending on the number of the controller's challenges. Lets assume the controller has sent three challenges $\{C_1, C_2, C_3\}$, then the third step has to be repeated three times as follows:

$$\beta = r \cdot \prod C_{1=1} X_1 \bmod n$$

$$\beta = r \cdot \prod C_{2=1} X_2 \bmod n$$

$$\beta = r \cdot \prod C_{3=1} X_3 \bmod n$$

Algorithm3: *Login and Identification Phase*

The host has to prove its identity to the controller

Log_Phase ()

Input:	*Random numbers C_j and r*
	Challenges $C_j = (C_1, ..., C_k)$
Output:	β
Procedure:	
Begin	{ *Host: H*
	calculate: $\lambda = \mp r^2 \bmod n$
	send: $C \leftarrow \lambda$
	}
	{ *Controller: C*
	create: $C_j = (C_1, ..., C_k)$
	send: $H \leftarrow C_j$
	}
	{ *Host: H*
	calculate: $\beta = r \cdot \prod C_{j=1} X_j \bmod n$
	}
	{
	send: $C \leftarrow \beta$
End	}

Algorithm4: *Verification Phase*

The controller has to verify the host proof of identity

Ver_Phase ()

Input:	β
	Permission for communication
Output:	
Procedure:	{ *Controller: C*
Begin	*Verify:* {
	if $(\lambda = \mp \beta^2 \cdot \prod C_{j=1} Y_j \bmod n)$
	send: permission for communication
	else
	reject communication
	}
End	

3.4 Verification Phase

In this phase, the controller has to verify the correctness of proof of identity that the host has sent. The verification process occurs as follows:

- The controller verifies the correctness of the host proof of identity by calculating $\lambda = \mp \beta^2 \cdot \prod C_{j=1} Y_j \bmod n$.

And as per the assumption in the previous Sect. 3.3, the host has to repeat the proof identity three times, the controller in this phase has to repeat the step above three times as follows:

$$\lambda = \mp \beta^2 \cdot \prod C_{1=1} Y_1 \, mod \, n$$

$$\lambda = \mp \beta^2 \cdot \prod C_{2=1} Y_2 \, mod \, n$$

$$\lambda = \mp \beta^2 \cdot \prod C_{3=1} Y_3 \, mod \, n$$

After repeating the verification step three times, if the host's proof of identity is correct, the host can get the network rules and policies and send and receive the network packets. If the host proof of identity is not correct, then the host must reattempt to log in again with the correct identity proof.

Table 2. Security goals of ZKPAUTH.

Goal	Description
secrecy_of sec_PW	Secrecy of host password
secrecy_of sec_NH	Secrecy of the nonce sent from NDC to the host
secrecy_of sec_NC	Secrecy of the nonce sent from NDC to the controller
authentication_on auth_NH	Authentication of the nonce sent from NDC to the host
authentication_on auth_NC	Authentication of the nonce sent from NDC to the controller
authentication_on auth_X	Authentication of the λ
Authentication_on auth_C	Authentication of the challenge C_i
authentication_on auth_Y	Authentication of the β

4 Implementation and Formal Security Verification

In this section, the implementation and the formal security verification of ZKPAUTH have been done using the Automation Validation of the Internet Security Protocol and Applications (AVISPA) tool [26]. It is a widely-used tool designed to simulate and verify the security of protocols proposed for security aspects and various security protocols have been verified using AVISPA [27–29]. The High-Level Protocol Specification Language (HLPSL) [30] is used to write the security protocols' specifications to simulate them. The tool uses the Dolev-Yao intruder model [31] to eavesdrop, intercept and modify the messages sent among the security protocol's communication parties by initiation replay and MITM attacks. AVISPA tool contains four back-ends (OFMC, Cl-AtSe, SATMC and TA4SP) that automatically integrate with the HLPSL to automatically analyse the security protocol.

4.1 The Specification and Simulation Result

The specification of the ZKPAUTH has been done using the HLPSL language. There are three basic roles (Host, NDC, and Controller). Every role has to send the others with the specified messages of four phases of the ZKPAUTH. Additionally, there are two more roles (Session and Environment) responsible for managing the communication sessions between basic roles. The session role contains the intruder's knowledge, which initiates the message interruption and modification. The environment role includes the security goals that aim to be achieved during the execution time of the ZKPAUTH. These security goals determine the safety and efficiency of the security protocol. Eight security goals have been specified in HLPSL code, three goals are for secrecy, and the other five goals are for authentication, the eight specified goals have been achieved and they are explained in Table 2.

Depending on the specification roles and security goals, the ZKPAUTH was simulated using two back-ends (OFMC and CL-AtSe) to verify its security against network attacks. The simulation result OFMC back-end is SAFE and demonstrates the security of ZKPAUTH against replay and MITM attack. And the simulation result using the CL-AtSe is also SAFE, and it also demonstrates the security of ZKPAUTH against replay and MITM attack. Figure 2 (a) and (b) show the simulation result of both back-ends, respectively.

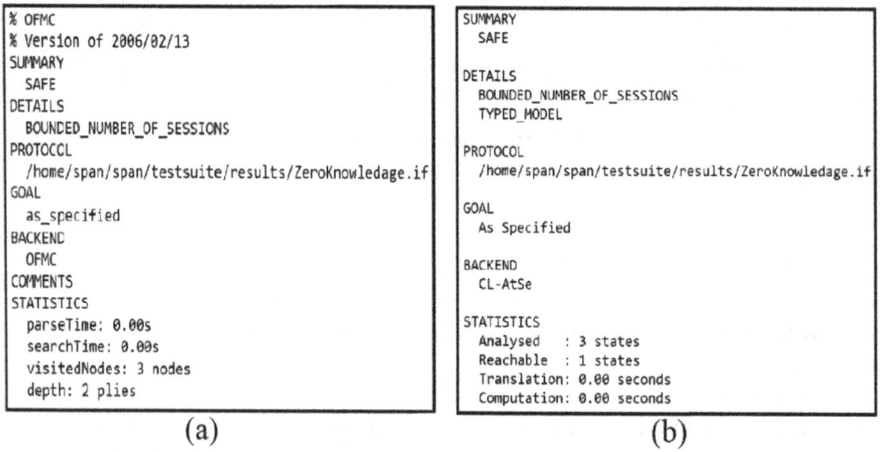

```
% OFMC
% Version of 2006/02/13
SUMMARY
  SAFE
DETAILS
  BOUNDED_NUMBER_OF_SESSIONS
PROTOCCL
  /home/span/span/testsuite/results/ZeroKnowledage.if
GOAL
  as_specified
BACKENC
  OFMC
COMMENTS
STATISTICS
  parseTime: 0.00s
  searchTime: 0.00s
  visitedNodes: 3 nodes
  depth: 2 plies
```

```
SUMMARY
  SAFE
DETAILS
  BOUNDED_NUMBER_OF_SESSIONS
  TYPED_MODEL
PROTOCOL
  /home/span/span/testsuite/results/ZeroKnowledage.if
GOAL
  As Specified
BACKEND
  CL-AtSe
STATISTICS
  Analysed   : 3 states
  Reachable  : 1 states
  Translation: 0.00 seconds
  Computation: 0.00 seconds
```

(a)   (b)

Fig. 2. (a) OFMC back-end simulation result of ZKPAUTH. (b) CL-AtSe back-end simulation result of ZKPAUT

5 Security Analysis

5.1 DoS Attack

The attacker will attempt to attack the controller to make it unavailable to the other network devices; thus, the hacker will try to login to the network illegally. ZKPAUTH is secure against the DoS attack because the controller is ordered to accept the hosts who only calculate the challenges C_i multiple times as it mentioned in the login phase as $\beta = r \cdot \prod C_{j=1} X_j \bmod n$.

5.2 MITM Attack

The MITM attacker fetches the messages getting sent between the host and the controller. In ZKPAUTH, the host sends its credentials $\{ID, h(PW), Y_j\}$ only once the time of registration. Furthermore, the host can log in into the network only if it calculates C_i's challenges and these challenges changed in every attempt. Thus, the MITM attack can't affect the ZKPAUTH.

5.3 Host Impersonation Attack

The attacker will try to impersonate the host's identity and attempts to login into SDN network. The host in ZKPAUTH has public and private keys where private keys are not published to other communication parties. Moreover, the host solves C_i's challenges using the private keys Xi where the attacker has no information about them. Thus, the attacker cannot impersonate the host identity and fool the controller.

5.4 Dictionary Attack

The dictionary attack tries to break the password by entering every word as a dictionary to gain a login to the network. In ZKPAUTH the dictionary attack cannot affect the network because it only uses the password in the registration phase. The host uses the nonce n to gain the login to the network, which the attacker, in this case, cannot get a solution of nonce as it is a complex mathematical operation.

5.5 Brute Force Attack

The Brute force attack works same dictionary attack with a difference of entering the words randomly using any possible combination of the words; this is faster than the dictionary attack but, in both cases, the ZKPAUTH is secure against these types of attacks because of the idea of ZKP that makes the nonce n the main parameter to make the login.

5.6 Mutual Authentication

ZKPAUTH ensures the mutual authentication between the host and controller, and that means the host has to trust the controller, and the controller has to trust the host. Mutual authentication occurs when the NDC sends the nonce n to both host and controller. In this case, the controller will trust the host if only and only if it solves the challenges, and the host trust the controller if the controller got the correct solution of the challenges.

6 Performance Evaluation

This section has compared the proposed scheme with other related schemes from the aspects of security features, computational cost, communication cost, and storage overhead. This comparison proves that the proposed scheme is more efficient than the existing schemes.

6.1 Security Features

ZKPAUTH was compared with other existing schemes and proved that ZKPAUTH is more secure than these existing schemes. The comparison is illustrated in Table 3 where the (Y) indicates that the scheme is supportive of the security feature, and (N) indicates that the scheme is non-supportive of this feature.

6.2 Computational Cost Communication Cost and Storage Overhead

In this section, ZKPAUTH was compared with other existing schemes. The comparison includes the aspects of computational cost communication cost and storage overhead. Without the loss of generality, the *ID, PW, Y, β, λ* is recommended to be 128-bit, *n* is recommended to be 512-bit, and *C* is recommended to be 256-bit. The notations T_e, T_{mod}, T_{ENC} T_H, T_{BIT}, T_M donate to the number of exponents, modulus, encryption-decryption process, hash function, bitwise operations and multiplication respectively. The mathematical operations of ZKPAUTH computational cost were counted and compared with other existing schemes. ZKP has no encryption-decryption operations and only one hash operation along with zero bitwise operation. These operations exhaust the processors, making the ZKPAUTH more efficient in computational cost than other existing schemes (see Table 4 and Fig. 3 (a).

Table 3. The comparison of the security features.

Security feature	Cho *et al.* [18]	Lee *et al.* [19]	Natanzi *et al.* [23]	ZKPAUTH
DoS attack	N	N	N	Y
MITM attack	Y	Y	Y	Y
Impersonation attack	N	N	N	Y
Replay attack	Y	Y	Y	Y
Dictionary attack	N	N	N	Y
Brute force attack	N	N	N	Y
Mutual authentication	Y	N	Y	Y

For the communication cost, the number of bits that must be transferred between communication parties was counted. Thus ZKPAUTH has fewer bits than other existing schemes, indicating the efficiency of ZKPAUTH see Table 5 and Fig. 3 (b). In storage overhead, the number of bits that every communication party has to store in its memory was counted. ZKPAUTH has to store fewer bits than other existing schemes, proving the effectiveness of ZKPAUTH over other existing schemes; see Table 6 and Fig. 3 (c).

Table 4. The comparison of the computational cost.

Scheme	T_e	T_{mod}	T_{ENC}	T_H	T_{BIT}	T_m
Cho et al. [18]	0	0	2	2	0	2
Lee et al. [19]	0	0	0	4	3	0
Natanzi et al. [23]	0	0	5	3	0	0
ZKPAUTH	4	3	0	1	0	4

Table 5. The comparison of the communication cost.

Scheme	Communication cost/bit
Cho et al. [18]	2320
Lee et al. [19]	3584
Natanzi et al. [23]	2688
ZKPAUTH	1920

Table 6. The comparison of the storage overhead.

Scheme	Storage overhead/bit
Cho et al. [18]	1664
Lee et al. [19]	2816
Natanzi et al. [23]	2048
ZKPAUTH	1664

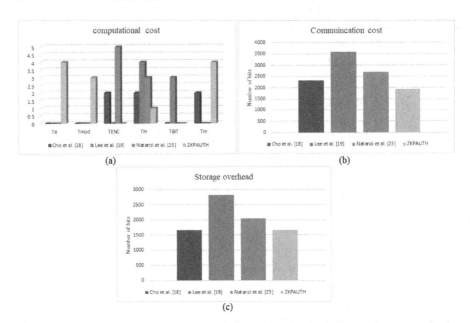

Fig. 3. (a) The comparison of the computational cost (b) the comparison of the communication cost (c) the comparison of the storage overhead.

7 Conclusion and Future Direction

As the communication between the controller and the hosts in SDN is not secure due to the complexity of TLS configuration in OpenFlow and not all network devices enforce the feature of TLS, SDN needs an effective authentication protocol to ensure security among the network. ZKPAUTH, an authentication scheme based on zero-knowledge proof was proposed to enforce the authentication between the controller and SDN hosts. The authentication occurs before establishing the communication between the devices to ensure the trust among them before they exchange the sensitive flow rules. ZKPAUTH security is tasted using the AVISPA tool that proves that it is safe against replay and MITM attacks. The security analysis which also proves the efficiency and immunity of ZKPAUTH against DoS and host impersonation attacks was demonstrated. The future work is to implement the proposed scheme in real SDN scenarios and extend the authentication process to be between the controller and OpenFlow switches.

References

1. Cao, Z., Member, S., Panwar, S.S., Kodialam, M., Lakshman, T.V.: Enhancing mobile networks with software defined networking and cloud computing. IEEE/ACM Trans. Netw. **25**(3), 1431–1444 (2017). https://doi.org/10.1109/TNET.2016.2638463
2. Xia, W., Wen, Y., Foh, C.H., Niyato, D., Xie, H.: A survey on software-defined networking. IEEE Commun. Surv. Tutorials **17**(1), 27–51 (2015). https://doi.org/10.1109/COMST.2014.2330903

3. McKeown, N., et al.: OpenFlow: enabling innovation in campus networks. ACM SIGCOMM Comput. Commun. Rev. **38**(2), 69–74 (2008). https://doi.org/10.1145/1355734.1355746

4. Oktian, Y.E., Lee, S.G., Lee, H.J., Lam, J.H.: Distributed SDN controller system: a survey on design choice. Comput. Netw. **121**, 100–111 (2017). https://doi.org/10.1016/j.comnet.2017.04.038

5. Krombholz, K., Mayer, W., Schmiedecker, M., Weippl, E.: 'I have no idea what I'm doing' – on the usability of deploying HTTPS. In: Proceedings of the 26th USENIX Security Symposium, pp. 1339–1356 (2017)

6. Fahl, S., Acar, Y., Perl, H., Smith, M.: Why Eve and mallory (also) love webmasters: a study on the root causes of SSL misconfigurations. In: ASIA CCS 2014 - Proceedings 9th ACM Symposium Information, Computer and Communications Security, pp. 507–512 (2014). https://doi.org/10.1145/2590296.2590341

7. Benton, K., Camp, L.J., Small, C.: OpenFlow vulnerability assessment. In: HotSDN 2013 – Proceedings of 2013 ACM SIGCOMM Workshop on Hot Topics in Software Defined Networking, pp. 151–152 (2013). https://doi.org/10.1145/2491185.2491222

8. Goldwasser, S.: The knowledge complexity of interactive proof systems. SIAM J. Comput. **18**(1), 108–128 (1989). https://doi.org/10.1090/psapm/038/1020812

9. Feige, U., Fiat, A., Shamir, A.: Zero-knowledge proofs of identity. J. Cryptol. **1**(2), 77–94 (1988). https://doi.org/10.1007/BF02351717

10. Abdullaziz, O.I., Wang, L.C.: Mitigating DoS attacks against SDN controller using information hiding. In: IEEE Wireless Communications and Networking Conference, WCNC, vol. 2019, pp. 1–6 (2019). https://doi.org/10.1109/WCNC.2019.8885764

11. Wang, M., Liu, J., Chen, J., Liu, X., Mao, J.: PERM-GUARD: authenticating the validity of flow rules in software defined networking. J. Signal Process. Syst. **86**(2–3), 157–173 (2016). https://doi.org/10.1007/s11265-016-1115-8

12. Ferrazani Mattos, D.M., Duarte, O.C.M.B.: AuthFlow: authentication and access control mechanism for software defined networking. Ann. Telecommun. **71**(11–12), 607–615 (2016). https://doi.org/10.1007/s12243-016-0505-z

13. Mutaher, H.: Openflow controller-based SDN: security issues and countermeasures. Int. J. Adv. Res. Comput. Sci. **9**(1), 765–769 (2018). https://doi.org/10.26483/ijarcs.v9i1.5498

14. Hong, S., Xu, L., Wang, H., Gu, G.: Poisoning network visibility in software-defined networks: new attacks and countermeasures. NDSS **15**, 8–11 (2015). https://doi.org/10.14722/ndss.2015.23283

15. Zhou, R., Lai, Y., Liu, Z., Liu, J.: Study on authentication protocol of SDN trusted domain. In: Proceedings - 2015 IEEE 12th International Symposium on Autonomous Decentralized Systems, ISADS 2015, pp. 281–284 (2015). https://doi.org/10.1109/ISADS.2015.29

16. Buragohain, C., Medhi, N.: FlowTrApp: an SDN based architecture for DDoS attack detection and mitigation in data centers. In: 3rd International Conference on Signal Processing and Integrated Networks, SPIN 2016, pp. 519–524 (2016). https://doi.org/10.1109/SPIN.2016.7566750

17. Cui, H., Chen, Z., Yu, L., Xie, K., Xia, Z.: Authentication mechanism for network applications in SDN environments. In: International Symposium on Wireless Personal Multimedia Communications, WPMC, vol. 2017, pp. 1–5 (2018). https://doi.org/10.1109/WPMC.2017.8301788

18. Cho, J.Y., Szyrkowiec, T.: Practical authentication and access control for software-defined networking over optical networks. In: SecSoN 2018 - Proceedings of the 2018 Workshop on Security in Softwarized Networks: Prospects and Challenges, Part SIGCOMM 2018, pp. 8–13 (2018). https://doi.org/10.1145/3229616.3229619

19. Lee, J., Park, M., Chung, T.: Path information based packet verification for authentication of SDN network manager. In: Park, J., Stojmenovic, I., Jeong, H., Yi, G. (eds.) Computer Science and its Applications. Lecture Notes in Electrical Engineering, vol. 330. Springer, Heidelberg (2015). https://doi.org/10.1007/978-3-662-45402-2_122

20. Ellinidou, S., et al.: SSPSoC: a secure SDN-based protocol over MPSoC. Secur. Commun. Netw. **2019**, 1–11 (2019). https://doi.org/10.1155/2019/4869167

21. Fang, L., et al.: THP: a novel authentication scheme to prevent multiple attacks in SDN-based IoT network. IEEE Internet Things J. **7**(7), 5745–5759 (2020). https://doi.org/10.1109/JIOT.2019.2944301

22. Major, W., Buchanan, W.J., Ahmad, J.: An authentication protocol based on chaos and zero knowledge proof. Nonlinear Dyn. **99**(4), 3065–3087 (2020). https://doi.org/10.1007/s11071-020-05463-3

23. Natanzi, S.B.H., Majma, M.R.: Secure distributed controllers in SDN based on ECC public key infrastructure. In: 2017 International Conference on Electrical and Computing Technologies and Applications, ICECTA 2017, vol. 2018, pp. 1–5 (2017). https://doi.org/10.1109/ICECTA.2017.8252015

24. Ambili, K.N., Jose, J.: A secure software defined networking based framework for IoT networks. J. Inf. Secur. Appl. **2020**, 1–19 (2020)

25. Shankaraiah, S.: Security and authentication scheme for software defined network. Int. J. Innov. Technol. Explor. Eng. **9**(4), 1116–1128 (2020). https://doi.org/10.35940/ijitee.d1659.029420

26. Armando, A., et al.: The AVISPA tool for the automated validation of Internet security protocols and applications. In: Etessami, K., Rajamani, S.K. (eds.) CAV 2005. LNCS, vol. 3576, pp. 281–285. Springer, Heidelberg (2005). https://doi.org/10.1007/11513988_27

27. Farash, M.S., Turkanović, M., Kumari, S., Hölbl, M.: An efficient user authentication and key agreement scheme for heterogeneous wireless sensor network tailored for the Internet of Things environment. Ad Hoc Netw. **36**, 152–176 (2016). https://doi.org/10.1016/j.adhoc.2015.05.014

28. Viganò, L.: Automated security protocol analysis with the AVISPA tool. Electron. Notes Theor. Comput. Sci. **155**(1), 61–86 (2006). https://doi.org/10.1016/j.entcs.2005.11.052

29. Salman, O., Abdallah, S., Elhajj, I. H., Chehab, A., Kayssi, A.: Identity-based authentication scheme for the Internet of Things. In: Proceedings - IEEE Symposium on Computers and Communication, vol. 2016, pp. 1109–1111 (2016). https://doi.org/10.1109/ISCC.2016.7543884

30. Chevalier, Y., et al.: A high level protocol specification language for industrial security-sensitive protocols to cite this version: HAL Id: inria-00099882 A High-Level Protocol Specification Language for Industrial Security-Sensitive Protocols ∗ (2006)

31. Dolev, D., Yao, A.C.: On the security of public key protocols. IEEE Trans. Inf. Theory **29**(2), 198–208 (1983). https://doi.org/10.1109/TIT.1983.1056650

Efficiency Evaluation of Handover Management Techniques in LTE Heterogeneous Networks

Manoj$^{(\boxtimes)}$ ⓘ and Sanjeev Kumar

Department of CSE, Guru Jambheshwar University of Science and Technology, Hisar, India

Abstract. Handover management has been considered major factors showing the effectiveness of each wireless network technology. In this paper, the research has considered the role of fuzzy-logic tactics and Q-learning in decision making for handover margin. The optimization technique PSO has to provide an optimized solution during HOM. The optimized HOM performance, A3-based traditional scheme, and A2-A4-RSRQ handover algorithm are discussed based on user speed and traffic load parameters. An issue is that traditional researches have not focused on optimization mechanism to optimize user speed and traffic load. Such models were unable to optimize variable user speeds as well as dynamic traffic loads. An efficient optimizer is required to reduce the call drop rates and increase system throughputs. It is required to integrate a PSO mechanism during HOM to optimize performance to reduce the call drop rate. The research work also concludes the best approach for handover in LTE Network.

Keywords: LTE · Handover · PSO · RSRP · RSRQ · UE parameters

1 Introduction

Handover has been considered one of the key procedures to ensure that users are moving through the network but he should stay connected and provide QoS. User satisfaction is the key indicator of success. Presently in the case of mobile networks, handover (HO) optimization has been performed manually. HOM is performed manually on a daily or weekly basis. But the issue is that this time of approach consumes a lot of time. Moreover, it might not be performed as often as required. The solution to such a problem is the self-optimizing mechanism that would tune parameters during the HOM process. The focus of such a mechanism is to boost overall network performance along with improvement in QoS. The major focus of such a proposal is to minimize the count of handovers that have been initiated but not executed for completion. These are also known as Handover failures. The recent topic of research in Handover margin is self-optimization [1]. Handover has been explained by the very precise flow of events. In our approach, the PSO has been used to optimise of HOM and its performance has been compared to traditional HOM mechanism that could be Fuzzy based, A3-based, and A2-A4-RSRQ handover algorithm. The mechanisms major challenge is to detect the optimized triggering point consider different influencing factors such as traffic load and user speed.

© Springer Nature Switzerland AG 2021
A. Solanki et al. (Eds.): AIS2C2 2021, CCIS 1434, pp. 121–134, 2021.
https://doi.org/10.1007/978-3-030-82322-1_9

1.1 Handover Technique

There have been various handover mechanisms such as RSRP dependent A3-RSRP as well as RSRQ dependent A2-A4-RSRQ. This research is considering fuzzy-based and PSO based HOM to get optimum solution. These mechanisms are providing better performance solution for decision making. The handover parameters that have been optimized using RSRP Algorithm are considered Hysteresis and Time-to-Trigger (TTT). On the other hand, RSRQ mechanism has been considered serving cell threshold and a Cell Offset Neighbor. However, best cost function is performed using PSO mechanism [2].

1.1.1 A3-RSRP

A3-RSRP algorithm is considering different variations of EU numbers. It is considering the speed of the EU along with the conditions of the channel. Operations are performed with as well as without fading.

TTT = 256, 320, 480, 512, 640 in millisecond.

Hysteresis = 1, 2, 3, 4, 5, 6, 7, 8, 9, 10, 11, 12, 13, 14, 15 (in dB).

1.1.2 A2-A4-RSRQ

A2-A4-RSRQ considers the different variations in case of the EU numbers. It considers EU movement speed along with channel conditions. Operations are performed with and without fading.

ServingCellThreshold = 28, 29, 30, 31, 32 (in dB).

NeighbourCellOffset = 1, 2, 3, 4, 5, 6, 7, 8, 9, 10, 11, 12, 13, 14, 15 (in dB).

1.1.3 Fuzzy Based

Fuzzy logic has been considered as a mechanism for calculation depending on degrees of facts instead of true/false. It works on 1 or 0 the Boolean logic where modern computer works. In other words fuzzy logic works based on values between 0 to 1. The fuzzy supports the decision making on random conditions for handover margin.

1.1.4 Particle Swarm Optimization (PSO)

It is the well-known optimization mechanism that uses the objective function, lower bound, and upper bound. The local best and global best is calculated in different iterations. The best solution is considered as the optimum solution that supports decision-making in HOM.

In Sect. 1, the HOM parameters and existing techniques used during the handover parameter optimization algorithm are explained. Section 2 explains the existing researches in the field of HOM. Section 3 is presenting the problem statement where issues in existing researches have been discussed. Section 4 is presenting the performance analysis to compare the optimized HOM to the traditional HOM mechanism. Finally, Sect. 5 is showing the conclusion of the research.

2   Literature Review

In this section, Authors have reviewed various existing techniques related to the LTE handover system and provide various approaches related to them. Authors have introduced optimization of Handover parameter in LTE networks that is self organized [1]. Such algorithms consider account weighting factor provided by operator policy at various metrics of performance like the ratio of failure, CDR, and ping pong ration during (Table 1).

Table 1. Summary from literature survey

S.No	Author/Year	Objective of research	Mechanism/Method used	Criteria	Results
1	Thoms Jansen 2010 [1]	Optimization of HOM in LTE to implement self-organizing networks	Research has used self-optimizing algorithm in to manage handover parameters of a LTE	HOM failure ratio, CDR ratio, ping pong handover ratio	Simulation outputs are showing that optimization mechanism has increased the performance of system significantly
2	Mahmoud Mandour 2019 [3]	Optimization of Handover and predicting the user Mobility Prediction	Mechanism to forecast suitable target FAP in case of HO	Load balance of target FAP as well as whole network	Results show that algorithm would increase as well as decrease the probability of HO success as well as failure
3	R. Ahmad 2017 [4]	Considering decision making algorithm during Handover in LTE	Decision Algorithm	LTE based wireless network	Efficient decision making for handover
4	S. H. S. Ariffin 2013 [5]	Predicting mobility by Markov model for LTE based femtocell	Markov model	LTE based femtocell network	Efficient mobility prediction
5	A. M. Miyim 2014 [6]	Proposing prediction for mobility prediction with the support of Markov model for LTE based femtocell	Markov model	LTE based femtocell network	Enhanced mobility predication
6	A. Ulvan 2013 [7]	Proposing Handover mechanism as well as decision making in case of LTE dependent femtocell network	Handover decision making	LTE dependent femtocell network	Mechanism played a significant role in handover decision making efficiently

(continued)

Table 1. (*continued*)

S.No	Author/Year	Objective of research	Mechanism/Method used	Criteria	Results
7	T. V. T. Duong 2012 [8]	Proposing efficient mechanism to perform prediction of mobility for wireless network	Temporal weighted mobility rule	Mobility prediction	Research provided an effective approach for mobility prediction
8	M. Z. Chowdhury 2011 [9]	Optimization of neighbor cell list in case of femtocell-to-femtocell handover in case of dense femto cellular networks	Neighbor cell list optimization mechanism	Dense network	The optimization of the neighbor cell list made effectively
9	H. SI, Y. WANG 2010 [10]	Prediction of mobility for cellular network with the help of hidden Markov model	hidden Markov model	Mobility in cellular network	The prediction of mobility made efficiently in a cellular network
10	H. Ge/2009 [11]	To propose prediction system for handover management in case of LTE based systems	History-based handover prediction	LTE	Handover prediction made rapidly
11	A. Ulvan 2009 [12]	Enhancing handover mechanism by predicting mobility during broadband wireless access	Handover strategy	Mobility in wireless network	The handover mechanism has been enhanced
12	Z. H, W. X, 2010 [13]	Proposing efficient handover mechanism among femtocell as well as macrocell in case of LTE dependent networks	Handover technique	LTE based network	The research has provided a novel handover approach for LTE network
13	K. T, Y. Q 2007 [14]	To perform Mobility management mechanism with support of handover prediction in case of third generation LTE dependent system	Mobility management technique	3g LTE system	Research is useful to manage mobility in 3g LTE system

(*continued*)

Table 1. (*continued*)

S.No	Author/Year	Objective of research	Mechanism/Method used	Criteria	Results
14	P. G, S. B, R. S 2014 [15]	To provide suitable mechanism during mobility management for LTE based femtocells	Approach to manage mobility	LTE based femtocell network	Research has provided a novel approach to managing the mobility
15	Y. Kirsal 2016 [16]	To implement analytical modeling for new handover mechanism in order to enhance resource allocation	Handover mechanism to manage the allocation of resource	Environment that is Highly mobile	Resources have been managed effectively

3 Problem Statement

In LTE network, during HOM it is observed that modifying user speeds and variable loads of traffic play a significant role in increasing the call drop rate that would reduce the network's performance. Moreover, call drop rate is also considered a challenging aspect that affects telecom networks' quality of services. However, there have been fuzzy logic tactics and Q learning approaches are in existence to a decided whether handover margins should be made or not. Moreover, there is the existence of traditional A3 based on HOM. Along with these mechanisms, there is the existence of the A2-A4-RSRQ handover algorithm. But the limitations of these mechanisms are performance and accuracy. It has been observed that there is a need to find the optimized user speed and traffic load to manage the call drop rate. Previous researches are not focusing on optimization mechanism to trace optimized user speed and traffic load. Previous research is not capable of proposing the mechanism that is capable of optimizing the variable user speeds and dynamic traffic loads. An efficient optimizer is capable of providing reduced call drop rates and increases system throughputs for excellent QoS. There is a requirement for PSO to propose an optimization mechanism to increase performance and accuracy during handover management as well as to minimize the call drop rate. HOM and TTT need to be optimized using PSO to achieve this objective.

4 Performance Metrics

The metrics which have been used during the handover parameter optimization algorithm have been categorized into 3 categories.

1. System metrics have been considered as Reference signal received power as well as signal-to-interference along with noise ratio. These have chosen connected cells as well as handover candidates that are possible.

2. Control parameters have been managed by an algorithm supporting optimization to boost the performance of handover on the network. Handover gets started when conditions are met. RSRP of a cell should be more as compare to RSRP of the connected cell with hysteresis value. It is holding at a time in the time-to-trigger parameter.

3. Assessment metrics have been utilized as measurements at the time of the optimization process and performance indicators in the case of evaluation of optimization mechanism.

4.1 Reference Signal Received Power (RSRP)

RSRP has been computed from cell transmit power known as P_c. Moreover, the operator's pathless data to various cells (L_{ue}) and shadow fading having log-normal distribution plays a role in RSRP calculation. RSRP also requires a standard deviation of 3 dB termed as L_{fad}. The output of RSRP is computed using Eq. 1:

$$RSRP_{c,ue} \rightarrow P_c - L_{ue} - L_{fad} \qquad (1)$$

Value of signal to interference is compared with help of RSRP of connected cell termed as $RSRP_{conn}$. Values of RSRP of the strongest interfering cells are added to thermal noise. Value of RSRP of interferers and thermal noise have been added up to $RSRP_{int,noise}$. This has been considered that all cells are transferring with 46 dBm on exact simulation time. Value of signal to interference is computed with the help of Eq. 2:

$$SINR_{ue} \rightarrow RSRP_{conn} - RSRP_{int,noise} \qquad (2)$$

4.2 Time to Trigger

TTT, also termed as triggering time in the case of LTE networks, has been specified in the case of 3GPP. TTT values could be 0, 0.04, 0.064, 0.08, 0.1, 0.128, 0.16, 0.256, 0.32, 0.48, 0.512, 0.64, 1.024, 1.280, 2.560, and 5.120. Such sixteen values have been considered valid time to trigger data. There has been three thirty six control parameter integration from hysteresis as well as the value of TTT.

4.3 Failure Ratio in Case of Handover

Such a ratio is also termed HPI_{HOF}. This has been the ratio of count in case of unsuccessful handovers also termed as NHO_{fail} with respect to the number of attempts during handover. The count of handover is attempting has been a sum of the count of successful termed as NHO_{succ} and the count of failed handovers.

4.4 Ping-Pong Handover Ratio

When the call is transferred to a recent cell and is transferred back to source cell in small during as compare to critical time known as T_{crit}. These handover has been known as ping-pong handover. Such a ratio is termed as HPI_{HPP}. It represents the count of ping-pong handovers known as NH_{Opp} is divided by the overall count of handovers. The count of ping-pong handovers is termed as N_{HOpp}. The count of handovers when no ping-pong is occurring is NHOnpp. Count of failed handovers has been using N_{HOfail}.

4.5 Call Dropping Ratio

It is also termed as HPI_{DC} that is known as probability where existing call gets dropped before completion. This occurs during handover when the user is shifting out from coverage.

5 Simulation Result Evaluation

The simulation work is performing a comparative analysis of speed in case of A3 Based, Fuzzy based, A2-A4 based, PSO based mechanism considering HOM, CDR, PING PONG, Handover failure, Handover rate. In simulation work, the various parameters taken into consideration are HOM, CDR, Ping pong, handover failure, and handover rate. The calculations are made considering coverage radius (2 km). The overlapping radius is 500 m of two base stations. The user speed (Vi) ranges between 0 to 140 at the interval of 10 ms. The value of constant K is 1000, and time t is 2 s (Table 2).

The HOM has been calculated using Eq. 1

$$HOM(i) = \log 10((r - (Vi * t))/(r + (Vi * t) - s)) * K \tag{3}$$

Whereas CDR has been calculated using Eq. 4 and 5

$$alpha = (2 \, X \, r)/(Vi \, X \, t) \tag{4}$$

$$CDR = (1 - \exp(-alpha) \, X \, (1 - alpha))/(2 \, X \, alpha) \tag{5}$$

Equation 4 is presenting handover failure ratio

$$HPI_{HOF} = N_{HOfail}/(N_{HOfail} + N_{HOsucc}) \tag{6}$$

Equation 5 is presenting ping pong ratio

$$HPI_{HPP} = N_{HOpp}/(N_{HOpp}) + (N_{HOnpp}) + (N_{HOfail}) \tag{7}$$

Considering Eq. 3–7, the simulation has been performed to compare the performance of HOM, CDR, PING-PONG, Handover failure, and Handover rate calculation using A3 Based, Fuzzy based, A2-A4 based, and PSO. After applying these mechanisms, each algorithm's time consumption has been presented in Tables 3, 4, 5, 6 and 7, and their corresponding figures are presented as Figs. 1, 2, 3, 4 and 5 for HOM, CDR, Ping pong, handover failure, handover rate, respectively.

Table 2. Simulation parameters

Parameter	Value
User speed (m/s)	0,10,20,30,40,50,60,70,80,90,100,110,120,130,140
Total distance	2 km
Simulation time	Distance divide/speed
Mobility algorithm	Constant velocity Mobility model
Handover algorithm	A3- RSRP, A2-A4-RSRQ, Fuzzy, PSO
Serving cell threshold	28, 29, 30, 31, 32 (in dB)
Neighbor cell offset	1, 2, 3, 4, 5, 6, 7, 8, 9, 10, 11, 12, 13, 14, 15 (in dB)
Hysteresis	1, 2, 3, 4, 5, 6, 7, 8, 9, 10, 11, 12, 13, 14, 15 (in dB)
Time to trigger	256, 320, 480, 512, 640 in millisecond

Table 3. Comparison of HOM time consumption for different user speeds

Iteration	User speed	A3 Based	Fuzzy based	A2–A4 based	PSO based
1	0	0.8654	0.517	0.6207	0.376
2	10	0.3581	0.293	0.218	0.078
3	20	0.4235	0.3227	0.2387	0.054
4	30	0.4612	0.259	0.2556	0.05
5	40	0.2807	0.185	0.1643	0.048
6	50	0.2642	0.2448	0.1581	0.052
7	60	0.4741	0.238	0.2626	0.051
8	70	0.2627	0.2768	0.1558	0.049
9	80	0.3387	0.313	0.2049	0.071
10	90	0.4962	0.2717	0.2746	0.053
11	100	0.2735	0.3055	0.1623	0.051
12	110	0.4037	0.3209	0.2269	0.05
13	120	0.3812	0.2309	0.2161	0.051
14	130	0.6662	0.5591	0.5071	0.348
15	140	0.6486	0.4822	0.4788	0.309

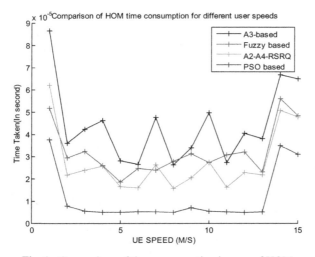

Fig. 1. Comparison of time consumption in case of HOM

Table 4. Comparison of CDR time consumption for different user speeds

Iteration	User speed	A3 Based	Fuzzy based	A2–A4 based	PSO based
1	0	0.2023	0.2042	0.1915	0.1808
2	10	0.0371	0.0215	0.0242	0.0114
3	20	0.0383	0.0345	0.0228	0.0073
4	30	0.0506	0.0303	0.0301	0.0097
5	40	0.0405	0.0286	0.0239	0.0072
6	50	0.0437	0.0254	0.0254	0.0071
7	60	0.0301	0.0375	0.019	0.0078
8	70	0.0592	0.0254	0.0333	0.0075
9	80	0.0397	0.0292	0.0235	0.0073
10	90	0.0321	0.0364	0.0203	0.0084
11	100	0.0626	0.0235	0.0348	0.0071
12	110	0.0552	0.0309	0.0315	0.0078
13	120	0.0334	0.0273	0.0206	0.0078
14	130	0.0631	0.0368	0.0355	0.0078
15	140	0.0611	0.0264	0.0344	0.0077

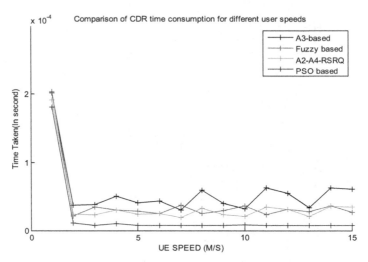

Fig. 2. Comparison of time consumption in case of call drop ratio

Table 5. Comparison of ping pong time consumption for different user speeds

Iteration	User speed	A3 Based	Fuzzy based	A2–A4 based	PSO based
1	0	0.4918	0.4561	0.4639	0.436
2	10	0.0997	0.0857	0.0838	0.068
3	20	0.086	0.0727	0.072	0.058
4	30	0.1178	0.0916	0.0959	0.074
5	40	0.1029	0.0734	0.0804	0.058
6	50	0.1102	0.1121	0.0981	0.086
7	60	0.1269	0.1056	0.104	0.081
8	70	0.1323	0.0945	0.1037	0.075
9	80	0.109	0.0769	0.087	0.065
10	90	0.0923	0.0857	0.0816	0.071
11	100	0.1147	0.0826	0.0878	0.061
12	110	0.1216	0.0891	0.0948	0.068
13	120	0.1119	0.0811	0.0894	0.067
14	130	0.0996	0.0815	0.0838	0.068
15	140	0.1141	0.0764	0.09	0.066

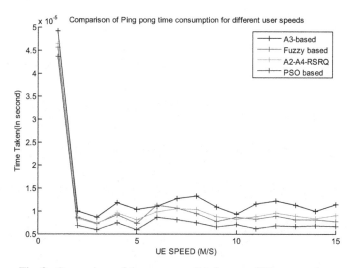

Fig. 3. Comparison of time consumption in case of Ping pong time

Table 6. Comparison of handover failure ratio time consumption for different user speeds

Iteration	User speed	A3 Based	Fuzzy based	A2-A4 based	PSO based
1	0	0.488	0.467	0.463	0.438
2	10	0.1257	0.1091	0.1048	0.084
3	20	0.0998	0.0856	0.0849	0.07
4	30	0.1465	0.1037	0.1173	0.088
5	40	0.1115	0.0846	0.0907	0.07
6	50	0.1054	0.0821	0.0867	0.068
7	60	0.115	0.0928	0.0975	0.08
8	70	0.1305	0.1039	0.1043	0.078
9	80	0.1106	0.0971	0.0898	0.069
10	90	0.1263	0.1044	0.1022	0.078
11	100	0.0948	0.0799	0.0819	0.069
12	110	0.1079	0.0927	0.0929	0.078
13	120	0.1084	0.0924	0.0937	0.079
14	130	0.1351	0.0925	0.106	0.077
15	140	0.1109	0.0959	0.0949	0.079

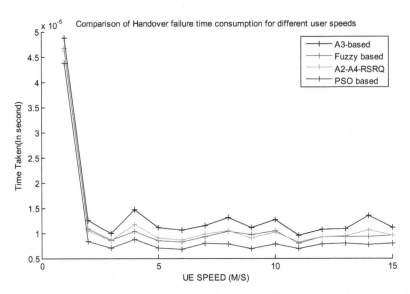

Fig. 4. Comparison of time consumption in case of handover failure ratio

Table 7. Comparison of handover failure ratio time consumption for different user speeds

Iteration	User speed	A3 Based	Fuzzy based	A2–A4 based	PSO based
1	0	1.028	0.7589	0.7776	0.527
2	10	0.283	0.3319	0.1768	0.071
3	20	0.474	0.303	0.2605	0.047
4	30	0.343	0.1572	0.1923	0.042
5	40	0.467	0.1677	0.2539	0.041
6	50	0.269	0.2492	0.1548	0.041
7	60	0.341	0.3387	0.1912	0.041
8	70	0.362	0.2051	0.2021	0.042
9	80	0.579	0.2764	0.3736	0.168
10	90	0.403	0.1912	0.2217	0.04
11	100	0.609	0.3305	0.3249	0.041
12	110	0.479	0.1662	0.2604	0.042
13	120	0.53	0.2139	0.2859	0.042
14	130	0.588	0.4887	0.4362	0.284
15	140	0.938	0.5946	0.6672	0.396

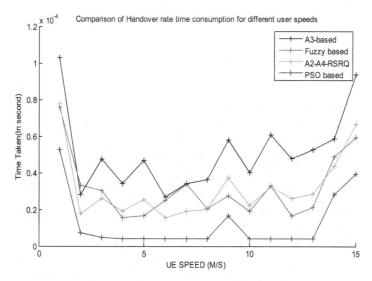

Fig. 5. Handover rate time consumption

6 Conclusion

The various handover approaches are implemented and observed that the user speed and traffic affect the call drop rate and handover margin. It is concluded that if the variable user speeds and variable traffic load increases then the call drop rate reduces networks' performance. PSO optimization mechanism supports finding the triggering point which is not possible with fuzzy logic or another traditional HOM mechanism such as A3-based tradition scheme, A2-A4-RSRQ handover algorithm. This technique has provided accurate solutions in minimum time. The research concludes that the PSO mechanism used to get the optimized result during simulation helps the optimized implementation of HOM with high performance. The research would play a significant role in finding the triggering point. Moreover, several other factors that influence HOM and CDR's performance could be considered in future research. This research would provide quality of service with high accuracy by reducing the call drop. The research would also allow the efficient management of LTE resources for better throughput. The PSO based handover optimization mechanism is changing values of hysteresis and TTT parameters in an automated fashion. Further, in future the PSO algorithm can be modified for even better performance of the network. Such algorithm consider the account weighting factor provided by operator policy to various performance metrics such as the ratio of handover failure, ping-pong handover, and call dropping.

References

1. Jansen, T., Balan, I., Turk, J., Moerman, I., Kurner, T.: Handover parameter optimization in LTE self-organizing networks. In: IEEE 72nd Vehicular Technology Conference, pp. 1–5 (2010)

2. Malekzadeh, M., Rezaiee, F.: Impact of inter-enodeB handover parameters on performance optimization in LTE networks. Indo. J. Electric. Eng. Comput. Sci. (IJEECS) **9**(1), 212–220 (2018)

3. Mandour, M., Gebali, F., Elbayoumy, A.D., Hamid, G.M.A., Abdelaziz, A.: Handover optimization and user mobility prediction in LTE femtocells network. In: IEEE International Conference on Consumer Electronics (ICCE), pp. 1–6 (2019)

4. Ahmad, R., Sundararajan, E.A., Othman, N.E., Ismail, M.: Handover in LTE-advanced wireless networks: state of art and survey of decision algorithm. Telecommun. Syst. **66**(3), 533–558 (2017). https://doi.org/10.1007/s11235-017-0303-6

5. Ariffin, S.H., Abd, N., Ghazali, N.E.: Mobility prediction via Markov model in LTE femtocell. Int. J. Comput. Appl. **65**(18) (2013)

6. Miyim, A.M., Ismail, M., Nordin, R.: Vertical handover solutions over LTE-advanced wireless networks: an overview. Wireless Pers. Commun. **77**(4), 3051–3079 (2014)

7. Ulvan, A., Bestak, R., Ulvan, M.: Handover procedure and decision strategy in LTE-based femtocell network. Telecommun. Syst. **52**(4), 2733–2748 (2013)

8. Duong, T.V.T., Tran, D.Q.: An effective approach for mobility prediction in wireless network based on temporal weighted mobility rule. Int. J. Comput, Sci. Telecommun. **3**(2), 29–36 (2012)

9. Chowdhury, M.Z., Bui, M.T., Jang, Y.M.: Neighbor cell list optimization for femtocell-to-femtocell handover in dense femtocellular networks. In: Third IEEE International Conference on Ubiquitous and Future Networks (ICUFN), pp. 241–245 (2011)

10. Si, H., Wang, Y., Yuan, J., Shan, X.: Mobility prediction in cellular network using hidden Markov model. In: 7th IEEE Consumer Communications and Networking Conference, pp. 1–5 (2010)

11. Ge, H., Wen, X., Zheng, W., Lu, Z., Wang, B.: A history-based handover prediction for LTE systems. In: IEEE International Symposium on Computer Network and Multimedia Technology, pp. 1–4 (2009)

12. Ulvan, A., Ulvan, M., Bestak, R.: The enhancement of handover strategy by mobility prediction in broadband wireless access. In: Proceedings of the Networking and Electronic Commerce Research Conference (NAEC 2009), pp. 266–276 (2009)

13. Zhang, H., Wen, X., Wang, B., Zheng, W., Sun, Y.: A novel handover mechanism between femtocell and macrocell for LTE based networks. In: 2nd IEEE International Conference on Communication Software and Networks, pp. 228–231 (2010)

14. Kim, T,H., Yang, Q., Lee, J.H., Park, S.G., Shin, Y.S.: A mobility management technique with simple handover prediction for 3G LTE systems. In: 66th IEEE Vehicular Technology Conference, pp. 259–263 (2007)

15. Ghosal, P., Barua, S., Subramanian, R., Xing, S., Sandrasegaran, K.: A novel approach for mobility management in LTE femtocells. Int. J. Wireless Mob. Net. **6**(5), 45 (2014)

16. Kirsal, Y.: Analytical modelling of a new handover algorithm to improve allocation of resources in highly mobile environments. Int. J. Comput. Commun. Control **11**(6), 789–803 (2016)

Advances Applications for Future Smart Cities

IoT Based Design of an Intelligent Light System Using CoAP

Ruchi Garg$^{(\boxtimes)}$ (iD) and Sonal Telang Chandel

Maulana Azad National Institute of Technology, Bhopal, Madhya Pradesh, India

Abstract. The Internet of things (IoT) is leading towards revolutionary applications with huge potential to improvise the efficiency of industries and environment multifold. Applications of IoT are marking their presence using the concept of a remote monitoring system, real-time data, visualization of data, and data analytics. This paper proposes and simulate an IoT based application of an Intelligent Light Control System. Luminosity sensors are used which sense the lux value. The application of light control system uses the lux value to control the switch ON/OFF of LED, remotely using the Constrained Application Protocol (CoAP protocol) at the application layer. The application is simulated in Cooja simulator along with Add-on plug-in Copper (Cu) in the Firefox browser. In this application, LEDs are switched ON/OFF automatically depending upon the availability of sunlight. The aim of this simulation is towards humungous power saving which will be a step forward to a green environment.

Keywords: Internet of Things · CoAP · 6LoWPAN · Wireless sensor network · Light control system · Contiki-Cooja · Copper (Cu)

1 Introduction

Internet-of-things (IoT) [1–3] is the need of time. People communicate and exchange information for multiple applications. Similarly machines need to talk to each other so they can work efficiently and productively. IoT play a vital role to make any device as "smart and connected device" [4–6]. Once connected to Internet, various sensors like temperature, humidity, motion, light, etc., are functional and operational parameters are captured and processed in the due applications. Internet-of-things has four stages 1) Data Acquisition - sensor layers which capture the data points 2) Communication Layer - send the data to the server for processing. 3) Data Process - the data capture is structured in backhand and format as required by applications 4) Applications Layer - the user interface where the data is visualized and decisions are taken. Dashboard is prepared on a real-time basis, so the user can act or improve his investment return. On the process data which is huge, later on the predictive, diagnostic and perspective analytics are run so the real data is meaningful. User can generate reports, predictions and simulate the future working. IoT is the lifeline for atomization and autonomous of the various products and applications.

© Springer Nature Switzerland AG 2021
A. Solanki et al. (Eds.): AIS2C2 2021, CCIS 1434, pp. 137–148, 2021.
https://doi.org/10.1007/978-3-030-82322-1_10

Gartner [7] in its yearly report specifies a significant rise in smart devices in the upcoming time almost 50 billion in the near future. In a connected network, every device must be assigned an IP address. This is necessary for Machine to Machine communication [8], which is the basis of IoT applications. IPv6 way of addressing up to 2^{128} unique addresses [9–11]. Hence IPv6 is considered over IPv4. IoT is a buzz word that covers our lives with applications in all sectors [12–18]. IoT concepts are now well accepted in various business domains like manufacturing, services, infrastructure, asset management, etc. [19]. They are widely used to improve efficiency [20], productivity, and analysis purposes. It helps in taking predictive and preventive steps. Efficiency enhancement and a green environment are their natural byproduct. This paper proposes a smart way to handle switching ON/OFF of a light remotely. The organization of the paper is as follows. A literature review related to work done in this field is presented in Sect. 2. The research methodology of the proposed work is explained in Sect. 3. The operating system Contiki, simulator Cooja, application layer protocol CoAP, Firefox plugin Copper are briefed in Sect. 4. The simulation environment setup and its details are given in Sect. 5. Implementation and demonstration screenshots of complete process flow are in Sect. 6. The conclusion is given in Sect. 7 followed by future work in Sect. 8.

2   Literature Review

Street lights are lit at a particular time every evening, no matter how much sunlight is there at that specific hour. Sometimes it gets dusk early and hence the need for lightening of street lights arises early. But it is not done as the time of switch on is fixed. The lighting system is for safe traffic and pedestrian movements and a better view of monuments and tourist attractions. A big amount of budget is spent on it. Not only the energy bill but also huge power consumption which should also be a matter of concern.

Y. Fujii et al. [21] have proposed a smart solution for streetlights that comprises various sensors and a communication network that could suffice a smaller distance. The brightness of LED is increased when there is traffic or pedestrians. It is decided from the input of the motion sensor. The proposed system consists of various things. It has an array of LEDs that have a system to adjust the power. The controller controls the system based on the sensing values of various sensors. On detection of motion, the controller unit turns on the LEDs.

The authors in [22] have used piezoelectric sensors in their proposed method. Using these sensors, the movement of the vehicles is detected in advance lights are switched on. When there is no vehicle, the lights are switched off.

P.T. Daely et al. [23] have used a Correlated Color Temperature (CCT) - centered radiance in the proposed system. A vital web server is designed in the system. This server works as a recipient of information of the weather conditions. Sensor-based streetlights having LEDs captures the data and send to the server. Conserving the energy is one advantage of this system. Besides this, the major impact is in reducing and avoiding accidents, which usually occurs because of poor visibility on roads.

The authors in research paper [24] have suggested a method centered on volume of the vehicular traffic. The proposed design is an intelligent method to control the lighting inside the tunnels. As mentioned in the paper, the system is implemented and tested in traffic situations of real-time. A number of vehicles and their speed is considered while doing the analysis.

Work is going on and various proposals are around [25–28]. In [26], the authors joined the idea of the Internet of Things (IoT) with LED enabled streetlights. The concept of microservices architecture has been implemented to streamline the development process and make it scalable based on demand. Incorporation of LED's hardware and Application Program Interface (API) is also described.

An advanced Wireless Sensor Network has been presented in [27] to monitor and control lighting arrangements of complex nature. In this system the strength and color temperature of the luminaires can be adjusted depending on sunlight. Especially built sensors calculate indoor lighting fluctuations in the initial process and extended for enough time to decide the correct control behaviour. The system not only adjusts the luminance but also observes various parameters of the exterior.

3   Research Methodology of Proposed Light Control System

As explained in Sect. 2, the idea of an intelligent light system is not novel. Still, resource constraint IoT devices for implementation of this system make the solution highly economical and hassle-free.

In this paper, a sensor-based intelligent lightning system is simulated. The intensity of available sunlight fully controls it. The sensors are connected to LEDs. Sensors will sense the lux value (luminosity). As soon as the lux value goes below a threshold, the LED is switched on.

On the other hand, the LED switches off when the lux value is above a threshold value. When lux value is above the threshold, this means that there is sufficient sunlight. There is no need for LEDs to be ON in presence of sufficient sunlight, hence they are switched off. A cut-off threshold value is decided and it is pushed in the software code. CoAP server having a light sensor is connected to the LED. All LEDs also work as actuators. The actuator means as soon as the sensor detects the lux value, it automatically switches ON/OFF the LED. CoAP client where the control application will run will be connected to the server through border router (BR). A response from the server will serve the client request for the value of sunlight. The complete scenario is shown in Fig. 1.

When there is a CoAP request in the proposed work, then the CoAP server sends the response. As shown in Fig. 1 there is a border router. The border router (BR) is on the edge of the 6LoWPAN. Its need is to connect the network to the outer world. In simulation, the border router connects the mote server with light sensors to the CoAP client running in the Firefox browser on the local machine.

As in Fig. 1, a bridge is to be created between the CoAP server and its client which is also CoAP enabled. It is done by enabling the Serial Socket Server of the mote set as Border Router (BR). For this, it uses the TunSLIP utility provided in Contiki. In Contiki, the Border Router Application uses the Serial Line Interface Protocol (SLIP). BR receives the global prefix through SLIP connection.

The CoAP client requests the value of sunlight (lux value). A response from the CoAP server serves it. Accordingly, the LED as an actuator will switch on/off. This will result in energy savings and reduction in operational cost compared to traditional lighting during fixed hours.

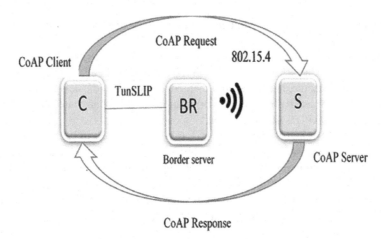

Fig. 1. Connectivity relation between components

4 Contiki-Cooja

Contiki is an operating system for the Internet of Things [29]. It is open-source and with the capability of multitasking. It is intended to be used for low-power microcontrollers with constraint resources. It provides low-power IP communication. It supports the fully standardized IPv4 and IPv6 and the recent light-weighted wireless standards such as 6LoWPAN [30–34], RPL, and CoAP. The software of Contiki is written in C Language. To use its applications, Ubuntu virtual machine is created using Contiki OS [35–39].

One of the network simulators used along with Contiki is Cooja [40, 41]. Simulation of a network having microcontroller nodes is done with the help of this simulator. Before programming the actual hardware microcontrollers, its simulation is made, tested, and analyzed. This makes Cooja, very suitable and convenient choice of the simulator for the Contiki network.

CoAP is a Constrained Application Protocol intended for the communication of IoT devices [42–45]. In this, resources of the server are available under a Unique Resource Locator (URL). Client nodes access those resources using procedures. These procedures are GET, PUT, POST, and DELETE in a generic browser like Copper (Cu). As stated in RFC 7228 [46], CoAP can handle microcontrollers of primary memory's very small capacity.

The Copper (Cu) CoAP user-agent is an additional feature available in the Firefox Web browser [47]. It is designed as an easy to use tool to manage the resources for the Internet of Things. The overall figure of networked embedded nodes in IoT is huge. To control and manage all the nodes is difficult for the users. Hence, Copper as a generic browser helps in managing and interacting with CoAP resources [48].

5 Simulation Environment

TelosB (TMote Sky) microcontrollers units are used, which are emulated as 6LoWPAN devices [49–53]. These microcontrollers units are also termed motes. These motes follow

IEEE 802.15.4. Their data rate is low usually 250 kbps and have 10kB RAM. One mote is set as Border Router and other motes as light sensors using the application programs in Contiki 2.7. The border router (BR) or edge router is a device on the edge of the 6LoWPAN network. Its need is to connect a network to the outside world. In simulation, the border router connects the mote server with light sensors to the CoAP client running in the Firefox browser on a local machine. The network topology of the simulation is shown in Fig. 2. It has three motes. Mote 1 is a border router and motes 2, 3 are servers having light sensors.

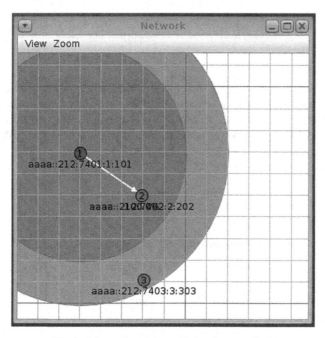

Fig. 2. Network topology (Color figure online)

The transmission range of motes is set to 50 m in diameter. The interference range is 70 m. The transmission range is seen in green color and the grey area represents the interference range of the particular mote. The motes are arranged fixedly. All the parameters set in the study are presented in Table 1.

Table 1. Simulation parameters

Parameter	Value
Operating system	Contiki 2.7
Browser	Firefox 51.0.1
Simulator	Cooja

(continued)

Table 1. (*continued*)

Parameter	Value
Plugin	Copper (Cu) 1.0.1
Packet analyzer	Wire Shark
Mote type	TelosB Sky motes
Mote placement model	Fixed
Transmission range	50 m
Interference range	70 m
Address type	Global IPv6 addressing

Protocols used in various layers are listed in Table 2. CoAP is a lightweight protocol. It is especially used in networks such as 6LoWPAN [54] which have limited resources. UDP is a preferred protocol at the transport layer to be used along with CoAP. RPL is a routing protocol for a low power network [55, 56]. It is a proactive distance vector protocol.

Table 2. Protocols used

Stack layer	Protocol used
Application layer	CoAP
Transport	UDP
Network	IPv6, RPL
Data link	802.15.4
Physical (radio type)	802.15.4

6 Implementation

A bridge is to be created between the simulated network (Fig. 2) and the local machine. This is done by enabling the Serial Socket Server of the mote set as Border Router (BR). For this, it uses the TunSLIP utility provided in Contiki [57, 58]. In Contiki, the Border Router Application uses the Serial Line Interface Protocol (SLIP). BR receives the global prefix through SLIP connection. BR set its global prefix and then it is communicated to the rest of the motes in the RPL network.

6.1 Connection of Border Router

When BR is connected, the global IPv6 addresses of all motes are set as

Mote 1 **aaaa : : 212:7401:1:101**
Mote 2 **aaaa : : 212:7402:2:202**
Mote 3 **aaaa : : 212:7403:3:303**

When the address of Border Router is entered in the web browser, it opens a page that displays all its neighbors and routes to all motes as shown in Fig. 3. Mote 3 is not in the BR (mote 1) range as shown in Fig. 2, but it will be connected through mote 2. A route is formed between BR and mote 3 as shown in Fig. 3. Hence all the communication between BR and mote 3 will be directed through mote 2.

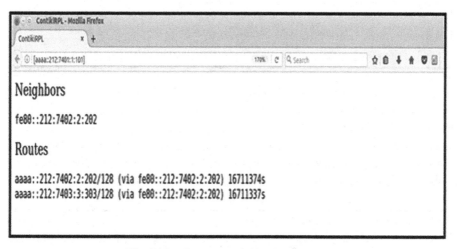

Fig. 3. Border router neighbors and routes

6.2 Response from a Light Sensor (Above Threshold Value)

The global IPv6 address (aaaa: : 212:7402:2:202) of mote 2 is entered in a browser and its resources are discovered. It has many resources, along with a light sensor. The sensor senses two values i.e. photosynthetic light and solar light. These values are fetched from it by using a GET command of CoAP in the web browser. The photosynthetic light and solar light values are displayed in the incoming tab's output window as shown in Fig. 4. A threshold value of luminosity as 100 is set. If the light sensor value goes above the threshold, then the LED remains off otherwise, it switches on.

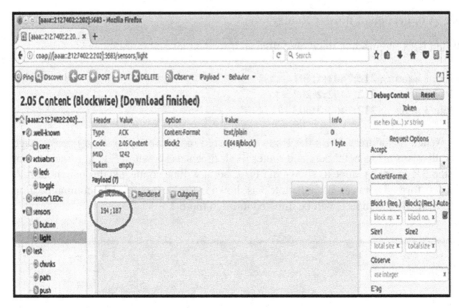

Fig. 4. Response from a light sensor

6.3   LED State is OFF

As visible in Fig. 4, the photosynthetic value of light is 194 and solar light is 187, which is higher than the preset threshold. As light value is high, which represents enough sunlight. Hence the LED remains OFF. The LED state is shown in Fig. 5.

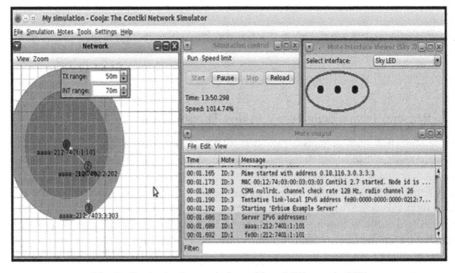

Fig. 5. Cooja simulator window with an LED state is OFF

6.4 Response from Light Sensor (Below Threshold Value)

A light sensor attached in the mote 2 keeps on sensing the sunlight. After a while, when sunlight becomes less and output of this, photosynthetic light and solar light are low. The values are 78 and 71, respectively as shown in the output window of an incoming tab of Fig. 6.

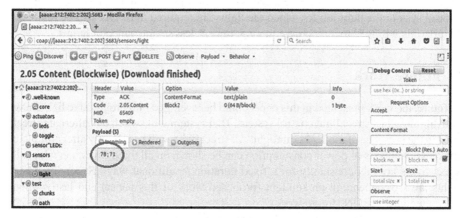

Fig. 6. Response from a light sensor

6.5 LED State is ON

Value sensed by the light sensor is less than the threshold, which indicates sunlight is less. Hence the red LED automatically gets switched on. This scenario is shown in Fig. 7. A similar task can be shown on Mote 3. In that case, the global IPv6 address of mote 3 is to be entered in the browser.

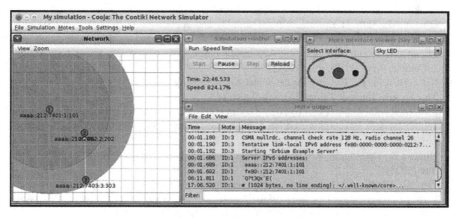

Fig. 7. Cooja simulator window with LED state is ON

7 Conclusion

In this paper, an Intelligent Light control system is proposed and simulated. All lights switch to an ON/OFF state based on the lux value of sensors present in the light. The work is an IoT solution using the CoAP protocol. The software part is written in the files of Contiki OS. It is simulated in the COOJA simulator. This application will result in huge power saving and result in reduced power consumption. In the presence of such a control system the duration for which the lights are required to be ON will be reduced. In turn, this will lead to an increase in glow time & less maintenance of lamps.

8 Future Work

In future, the work presented in this paper could be leveraged into progressive lightening and dimming in a street light control system. That system will switch ON alternate street light in the presence of moderate sunlight. All street lights will be ON only after complete darkness. Analysis of power consumption can be shown on all three cases, i.e. when all street lights are ON constantly for a fixed duration (traditional way). Secondly, when lights are ON depending on sunlight (proposed work of this paper) and lastly when alternate lights are ON (future work).

Also, motion sensors can be used provide additional features in the system. With such sensors' help, light system will be controlled by considering pedestrians and vehicles' movement.

References

1. Ashton, K.: That 'Internet of Things' thing. RFID J. **22**(7), 97–114 (2009)
2. Karimi, K., Atkinson, G.: What the Internet of Things (IoT) needs to become a reality. White Paper, FreeScale and ARM, pp. 1–16 (June 2013)
3. Feller, G.: The Internet of Things: In a Connected World of Smart Objects. Accenture & Bankinter Foundation of Innovation, pp. 24–29 (2011)
4. Own, C.M., Shin, H.Y., Teng, C.Y.: The study and application of the IoT in pet systems, pp. 1–8 (2013)
5. Maraiya, K., Kant, K., Gupta, N.: Application based study on wireless sensor network. Int. J. Comput. Appl. **21**(8), 9–15 (2011)
6. Ning, H., Liu, H.: Cyber-physical-social based security architecture for future Internet of Things. Adv. Internet Things **2**(01), 1–7 (2012)
7. Meulen, R.V.D.: Gartner says by 2020, a quarter billion connected vehicles will enable new in-vehicle services and automated driving capabilities. Gartner. STAMFORD Conn. (November 2015)
8. Hinden, R., Deering, S.: RFC3513: Internet Protocol Version 6 (IPv6) Addressing Architecture (2003)
9. Garg, R., Sharma, S.: Comparative study on techniques of IPv6 header compression in 6LoWPAN. In: Proceedings of the International Conference on Advances in Information Processing and Communication Technology (IPCT), Rome, Italy, pp. 34–38 (2016)
10. Forouzan, B.A.: Network Layer: Internet Protocol in Data Communication & Networking. 4th edn., pp. 582–597. McGraw-Hill, New York (2007). Ch. 20

11. Stalling, W.: Wireless Communication and Networks, 4th edn., pp. 39–118. Pearson Publication Limited, London (2004)
12. Wang, X., Song, X.: New medical image fusion approach with coding based on SCD in wireless sensor network. J. Electr. Eng. Technol. **10**(6), 2384–2392 (2015)
13. Zhang, D., Li, W., Liu, S., Zhang, X.: Novel fusion computing method for bio-medical image of WSN based on spherical coordinate. J. Vibro Eng. **18**(1), 522–538 (2016)
14. Zhang, X.: Design and implementation of embedded uninterruptible power supply system for web-based mobile application. Enterp. Inf. Syst. **6**(4), 473–489 (2012)
15. Chen, J., Mao, G.: Capacity of cooperative vehicular networks with infrastructure support: multi-user case. IEEE Trans. Veh. Technol. **67**(2), 1546–1560 (2018)
16. Zhang, D., Ge, H.: New multi-hop clustering algorithm for vehicular ad hoc networks. IEEE Trans. Intell. Transp. Syst. **20**, 1517–1530 (2018)
17. Zhao, C.P.: A new medium access control protocol based on perceived data reliability and spatial correlation in wireless sensor network. Comput. Electr. Eng. **38**(3), 694–702 (2012)
18. Zhang, D., Kang, X., Wang, J.: A novel image de-noising method based on spherical coordinates system. EURASIP J. Adv. Signal Process. **2012**, 1–10 (2012). https://doi.org/10.1186/1687-6180-2012-110
19. Evans, D.: The Internet of Things: How the next evolution of the Internet is changing everything. CISCO white paper, p. 1 (2011)
20. Garg, R., Sharma, S.: Cooja based approach for estimation and enhancement of lifetime of 6LoWPAN environment. Int. J. Sens. Wirel. Commun. Control **9**, 1–10 (2019)
21. Fujii, Y., Yoshiura, N., Takita, A., Ohta, N.: Smart street light system with energy saving function based on the sensor network. In: Proceedings of the Fourth International Conference on Future Energy Systems, pp. 271–272 (2013)
22. Abinaya, R., Varsha, V., Hariharan, K.: An intelligent street light system based on piezoelectric sensor networks. In: 2017 4th International Conference on Electronics and Communication Systems (ICECS), pp. 138–142 (2017)
23. Daely, P.T., Reda, H.T., Satrya, G.B., Kim, J.W., Shin, S.Y.: Design of smart LED streetlight system for smart city with web-based management system. IEEE Sens. J. **17**(18), 6100–6110 (2017)
24. Qin, L., Dong, L.L., Xu, W.H., Zhang, L.D., Leon, A.S.: An intelligent luminance control method for tunnel lighting based on traffic volume. Sustainability **9**(12), 2208 (2017)
25. Gharaibeh, A., et al.: Smart cities: a survey on data management, security, and enabling technologies. IEEE Commun. Surv. Tutor. **19**(4), 2456–2501 (2017)
26. Satrya, G.B., Reda, H.T., Woo, K.J., Daely, P.T., Shin, S.Y., Chae, S.: IoT and public weather data based monitoring & control software development for variable color temperature LED street lights. Int. J. Adv. Sci. Eng. Inf. Technol. **7**(2), 366–372 (2017)
27. Pierleoni, P., et al.: The scrovegni chapel moves into the future: an innovative Internet of Things solution brings new light to Giotto's masterpiece. IEEE Sens. J. **18**(18), 7681–7696 (2018)
28. Chang, Y.C., Lai, Y.H.: Campus edge computing network based on IoT street lighting nodes. IEEE Syst. J. **14**, 164–171 (2018)
29. Dunkels, A., Gronvall, B., Voigt, T.: Contiki-a lightweight and flexible operating system for tiny networked sensors. In: 29th Annual IEEE International Conference on Local Computer Networks, pp. 455–462 (2004)
30. Kim, E., Kaspar, D., Vasseur, J.: Design and application spaces for IPv6 over low-power wireless personal area networks (6LoWPANs). RFC6568 (2012)
31. Garg, R., Sharma, S.: Modified and improved IPv6 header compression (MIHC) scheme for 6LoWPAN. Wirel. Pers. Commun. **103**(3), 2019–2033 (2018)
32. Ismail, N.H.A., Hassan, R., Ghazali, K.W.M.: A study on protocol stack in 6LoWPAN model. J. Theor. Appl. Inf. Technol. (JATIT) **41**(2), 220–229 (2012)

33. Hui, J.W., Culler, D.E.: Extending IP to low-power, wireless personal area networks. IEEE Internet Comput. **12**(4), 37–45 (2008)
34. Culler, D., Chakrabarti, S.: Infusion IP. 6LoWPAN: Incorporating IEEE 802.15. 4 into the IP architecture. White paper (January 2009)
35. The Contiki OS. http://www.contiki-os.org/p/about-contiki.html. Accessed 13 Sep 2019
36. Contiki - Connecting the Next Billion Devices. www.sics.se/contiki/. Accessed 18 Nov 2019
37. The Contiki Operating System Documentation. http://www.sics.se/adam/contiki/docs/. Accessed 26 Nov 2019
38. ANRG Installation. http://anrg.usc.edu/contiki/index.php/Installation. Accessed 17 Oct 2019
39. ANRG. Contiki Tutorials. http://anrg.usc.edu/contiki/index.php/Contiki_tutorials. Accessed 18 Aug 2019
40. Gonizzi, P., Duquennoy, S.: Hands on Contiki OS and Cooja Simulator: Internet of Things and Smart Cities, pp. 1–15 (2013)
41. GitHub. An introduction to Cooja. https://github.com/contiki-os/contiki/wiki/An-Introduct ion-to-Cooja#The_COOJA_Simulator. Accessed 18 Nov 2019
42. Hartke, K., Shelby, Z.: Observing resources in coap. The Internet Engineering Task Force (IETF) draft-ietf-core-observe-02 (work in progress) (March 2011)
43. Shelby, Z., Hartke, K., Bormann, C., Frank, B.: constrained application protocol (CoAP). The Internet Engineering Task Force (IETF) draft-ietf-core-coap-06 (2011)
44. Gorrieri, A., Davoli, L., Picone, M.: Hands on CoAP: Exercises. Internet of things course (May 2015)
45. Shelby, Z., Hartke, K., Bormann, C.: Constrained application protocol (CoAP): draft-ietf-core-coap-13, IETF Trust (2012)
46. Bormann, C., Ersue, M., Keranen, A.: Terminology for constrained-node networks. Internet Engineering Task Force (IETF): Fremont, CA, USA, pp. 2070–1721 (2014)
47. Kovatsch, M.: Demo abstract: human-coap interaction with copper. In: 2011 International Conference on Distributed Computing in Sensor Systems and Workshops (DCOSS 2011), pp. 1–2 (2011)
48. Contiki-Copper. http://people.inf.ethz.ch/mkovatsc/copper.php. Accessed 15 Sep 2019
49. Montenegro, G., Kushalnagar, N., Hui, J., Culler, D.: Transmission of IPv6 packets over IEEE 802.15.4 networks. Internet proposed standard RFC, vol. 4944, p. 130 (2007)
50. Olsson, J.: 6LoWPAN demystified. Texas Instruments (2014)
51. Mulligan, G.: The 6LoWPAN architecture. In: Proceedings of the 4th Workshop on Embedded Networked Sensors, pp. 78–82 (2007)
52. Kim, E., Kaspar, D., Gomez, C., Bormann, C.: Problem statement and requirements for IPv6 over low-power wireless personal area network (6LoWPAN) routing. RFC (May 2012)
53. Garg, R., Sharma, S.: A study on need of adaptation layer in 6LoWPAN protocol stack. Int. J. Wirel. Microw. Technol. (IJWMT) **7**(3), 49–57 (2017)
54. Shelby, Z., Bormann, C.: Introduction in 6LoWPAN: The Wireless Embedded Internet, 1st edn., pp. 3–11. Wiley, United Kingdom (2009)
55. Winter, T., et al.: RPL: IPv6 routing protocol for low-power and lossy networks. rfc **6550**, 1–57 (2012)
56. Vasseur, J., Agarwal, N., Hui, J., Shelby, Z., Bertrand, P., Chauvenet, C.: RPL: the IP routing protocol designed for low power and lossy networks. Internet Protocol for Smart Objects (IPSO) Alliance, vol. 36 (2011)
57. Tunslip Utility. http://anrg.usc.edu/contiki/index.php/RPL_Border_Router#Tunslip_utility. Accessed 25 Oct 2019
58. Building Contiki's tunslip6. https://www.iot-lab.info/tutorials/build-tunslip6. Accessed 15 Oct 2019

Implementation of Touch-Less Input Recognition Using Convex Hull Segmentation and Bitwise AND Approach

A. Anitha[1]([⊠]) [iD], Saurabh Vaid[2] [iD], and Chhavi Dixit[2] [iD]

[1] School of Information Technology and Engineering,
Vellore Institute of Technology, Vellore, Tamil Nadu, India
aanitha@vit.ac.in
[2] School of Computer Science and Engineering,
Vellore Institute of Technology, Vellore, Tamil Nadu, India
{saurabh.vaid2018,chhavi.dixit2018}@vitstudent.ac.in

Abstract. Pandemic situations lead to a heightened risk of viral transmission, particularly in public spaces, that hampers humans' daily lives. Some of the general public's utilities such as vending machines, ATMs, etc. have the potential to quickly become viral hotspots and transmit diseases at unmatched and uncontrollable levels. A possible solution to solving such issue is to utilise the computer vision technique, to convert visual cues into the input to the device using hand gestures. The proposed technique can be implemented using webcams since it is a cost-effective and easy way to integrate hand gestures. The process of gesture recognition includes segregating the videos into frames. An effort has been taken to propose a hand gesture recognition using Convex Hull Segmentation and Bitwise AND approach. Finally, the proposed method demonstrate the viability of the proposed system based on its memory usage in different test environments.

Keywords: Computer vision · Artificial intelligence · Gesture recognition · System control · Human-computer interaction

1 Introduction

Pandemics give rise to concerns in the minds of people over shared contact. In places like the ATM and automated ticket vending machines in metro stations, keypads/touch screens are utilized for input. It is hard to keep such devices sanitized when it is are so frequently used. To evade keypad as input, other forms of input like audio and visual input utilisation is suggested. Webcams present a frugal hardware solution that can be facilely incorporated with already subsisting systems, and are already built-in in a majority of personal contrivances. The issue of using audio as input is that it can be easily affected by external disturbances. In such cases, the best alternative is a visual input utilizing the contrivance camera or an external camera, with live image processing.

The paper aims to engender a program for touchless input for system controls in sundry specialized domains, utilizing video input from an internal/externally connected

© Springer Nature Switzerland AG 2021
A. Solanki et al. (Eds.): AIS2C2 2021, CCIS 1434, pp. 149–161, 2021.
https://doi.org/10.1007/978-3-030-82322-1_11

webcam. Aside from having the advantage of contactless utilization, visual input withal can be more accessible to people who are not acclimated with the utilization of sundry contrivances. Thus, it can be considered a subsidiary in various scenarios, such as personal home use and public machines like kiosks, ATMs, etc. Thus, the paper explores the use of device webcams for taking in real time input of our hands and using the input from various domains of system control.

2 Literature Review

A fast algorithm for hand gesture recognition was proposed by Asnaterabi Malima et al. [1] in their research. The researchers offered hand gesture recognition and image processing and considered a fixed set of manual commands in a reasonably structured environment and developed a simple and effective gesture recognition approach. The general overview of steps states segmentation, followed by locating fingers and classifying those. It is proved to be invariant to translation, rotation and scale of hand within the region of interest. Another novel way of using real-time hand gesture recognition was proposed by Karthik Karkera et al. [2]. Hand gesture recognition was tested to change the direction movement of a bot. Initially, the gestures were converted to commands using image processing and then sent to a wireless robot to change its direction. The algorithm proposed got up to 80% accuracy in bright light and 100% in the shade. Sohom Mukherjee [3] et al. suggested in their paper a way for fingertip detection and tracking for air writing, without the aid of any hand-held devices. The researchers addressed the challenges faced while recognizing fingertips due to the small fingertip dimension and absence of any standard delimiting criterion. Also, reseachers used a R-CNN framework in their algorithm for hand detection, followed by segmentation and finger counting using geometrical properties of hand. The finger detection algorithm used a new signature function called distance-weighted curvature entropy.

A novel method for finger detection was proposed by Amrita Biswas [4]. The researcher addressed the challenges faced in the identification of number and coordinates of a number of fingers. The proposed solution used Hough transform for fingertips detection and searching for long Hough lines in the vicinity of identified coordinates for cross-verification. The algorithm was tested with hand posing at multiple angles and worked efficiently. In 2018, Hao Tang et al. [5] attempted to introduce a fast and robust hand recognition technique by using image entropy and image density clustering to exploit the keyframes from hand gesture video. It seemed to improve the efficiency of hand recognition. A feature fusion strategy was also proposed for increasing the recognition performance by improving feature representation. For testing their proposed algorithms, two datasets were introduced, and it was shown that the performance of the proposed algorithm was at par with other hand gesture datasets from Northwestern University, Cambridge. V. Harini et al. [6] built a Linux platform-based paper, compatible with Windows, used a fundamental algorithm requiring OpenCV and Numpy. Once the hand was isolated using thresholding and segmentation, a histogram was created for it. The histogram was created by tracking the hand of the user's hand over 9 boxes divided in a 3 * 3 grid. The histogram was then used to count fingers using the algorithm as mentioned in the paper by Asnaterabi Malima et al. A limitation was the background's specificity; the program's accuracy reduced when there were shadows or air gaps. Xing et al. [7] addressed the issue of loss of information while extracting features using classical methods. The proposed deep learning to improve accuracy of EMG-based hand gesture recognition with a parallel architecture with five convolution layers.

In 2019, Lalit Kane et al. [8] proposed a framework for a system capable of recognizing continuous dynamic gestures characterized by short-duration posture sequences by modifying the shape-matrix for hand silhouettes ans used the nearest neighbour and Naïve Bayes classifier with a windowing mechanism by acheiving mean accuracy of 95.2% with their approach. Shanthakumar et al. [9] proposed an angular-velocity based method directly applied to real-time 3D motion data streamed by a sensor-based system that could assess both static and dynamic gestures. Their approach showed high recognition accuracy, high execution performance, and high-levels of usability. Suja Palaniswamy et al. [10] discussed developing a wearable device for capturing hand gestures using accelerometers and gyroscopes that were able to capture four different hand gestures using their device and recognize those using the Support Vector Machine Algorithm. Benitez-Garcia G et al. [11] addressed real-time performance limitation due to intense extra computing cost because of optical flow by employing light-weight semantic segmentation method (FASSD-Net) to boost the accuracy of the TSN and TSM methods, gained significant accuracy over the classical methods. Santoshi G. et al. [12] developed a model for using one or more fingers for hand gesture recognition and performing functions like switch tabs, zooming in and out, and swiping by building a multilayer convolution neural network that considered pixel-level segmentation and detected a number of fingers and movements. Daniel M.S. et al. [13] suggested a framework for hand gesture recognition and conversion to text and speech to help dumb people communicate. The researcher used Keras as a platform, a webcam for real-time image capturing, and CNN for training the model. Runwal R. et al. [14] used ultrasonic sensors to track hand using sound waves' movement for controlling laptop features. The external hardware used were low-cost sensors for capturing movement and an Arduino board for translating to the proper input by using PyAutoGUI and PySerial for the final stage of processing and action on the laptop. Sangeeta Kumari et al. [15] implemented a hand gesture recognition based human-computer interaction system using tensor-flow. Their approach manipulated both static and dynamic hand gestures and were able to achieve an accuracy of 94.44% with their dataset.

In 2019, Md. Manik Ahmed et al. [16] proposed and compared four different hand gesture recognition systems and applied minor optimizations to improve existing models' accuracy. The final accuracy achieved was 93.21%, and the run time was 224 s. The optimizations applied could be used in other papers for improving their efficiency. Lizhi Zhang et al. [17] presented a Hidden Markov Model-based algorithm. It used two shape features for parameter training and identifying gesture categories hierarchically. The wavelet texture energy feature was used to reflect the internal details of the gesture image. The proposed method had a good recognition effect for gestures. Duy-Linh Nguyen et al. [18] used convolution layers, CReLU module and max-pooling layers alternatively for feature extraction and used two-sibling convolution layers for classification and regression as a detection block and achieved 93.32% average precision. Anitha [19] proposed cyber security method using IOT, so the proposed technique can be used for human-computer interaction.

Adam et al. [20] suggested touchless input for human-computer interaction aiming specifically at disabled people by exploing four types of touchless inputs for the disabled, namely, eye-tracking, brain-computer interfaces, speech recognition and computer vision approaches. Eye-tracking used pupil tracking and computer vision-based approaches used visible or infrared lights, both requiring webcams. A virtual keyboard, typing according to pupil movement, was demonstrated in the research. In 2017, ABD Albary et al. [21] suggested using fingers for pointing, targeted for mute people, making communication easier, suggested a novel way of drawing gestures in air in front of a camera, which can use image processing to translate it to meaningful speech or text. The researcher proposed extraction of only hand using the YCbCr color space filter and converting it to black and white image for counting of fingers. The image was used to count the number of black and white flips on a specific pre-defined path on the image. The number of flips translated to the number of fingers. acquired an accuracy of up to 98%. Oinam Rabita Chanu et al. [22] carried out a comparative study for vision-based and data-based hand gesture recognition technique and compared a static vision-based hand recognition algorithm, a real-time hand gesture recognition technique and a data glove-based technique. When tested under different conditions on ten different subjects, it showed that vision-based hand recognition techniques gave higher accuracy in bright light than data-glove-based hand recognition techniques. Yaida Forouton et al. [23] designed a user interface to control computer cursor by hand detection and gesture classification used a dataset with 6720 samples and four classes for that purpose. A CNN framework was trained on the the collected dataset to predict the labels and work accordingly. The proposed algorithm was able to achieve 91.88% accuracy under various backgrounds. Pramod Pisharady et al. [24] used a Bayesian model of visual attention to generate a saliency map for identifying the hand region in any image input and feature-based visual attention using high-level and low-level features and classified the resultant data using Support Vector Machine classifier. The proposed methodology had an accuracy of 94.36%. Priyanka R. et. al [25] presented a real-time system for hand gesture recognition based on the detection of features like orientation, the centre of mass, status of fingers, and their respective location in the image by implementing methodology of recognised gestures for opening frequently visited pages on websites.

The paper has been organized as follows: Sect. 1 starts with the introduction, followed by Sect. 2 as the literature review. The proposed methodology was explained in Sect. 3 followed by Experimental analysis and results as Sect. 4. Section 5 discussed the results and discussions and concluded with Sect. 6.

3   Proposed Methodology

The hand gesture recognition algorithm was based on the research done by Asnaterabi Malima et al. OpenCV is used to read the feed from the integrated webcam frame by frame. The first 30 frames are used for background consolidation, and afterwards, the foreground mask is taken. The display is divided into two regions-of-interest. The leftmost region acts as a menu selection utility, reading the fingertip as a cursor, and selects which feature the user wishes to use. The central region-of-interest is the main area that is used for gesture recognition. Figure 1. depicts the primary user interface with menu and segmentation region of interest.

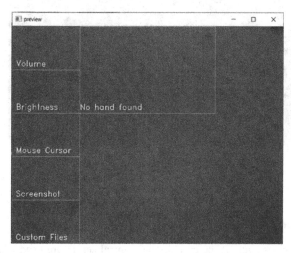

Fig. 1. Basic UI with menu and segmentation region of interests (ROIs)

The gesture recognition algorithm proposes taking the largest contour (assumed to be the hand) and calculating its center based on the extreme points. A tentative radius from the length and breadth of the convex hull formed so. Using Bitwise-AND, the software calculates where the circle is, and the contours of the hand intersect, and the quantity of these intersections are counted, thus enabling the system to read the number of fingers the user is holding up. Figure 2 and Figure 3 details the finger threshold made via foreground mask separation and finger contours.

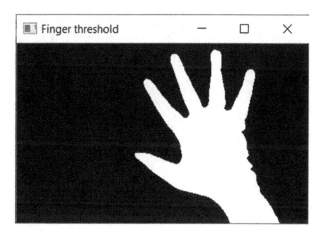

Fig. 2. Finger threshold for 5 fingers made via foreground mask separation

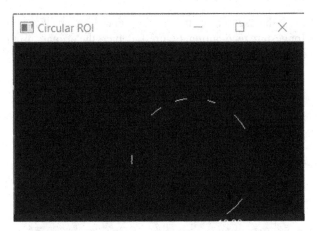

Fig. 3. Bitwise AND of finger contours and circle

3.1 Hardware and Software Specifications

The hardware devices such as the camera and central server are used. Camera must be an internal component of the system or externally connected to the program's system to function. If the proposed method is to be run on an eternal server, network architecture

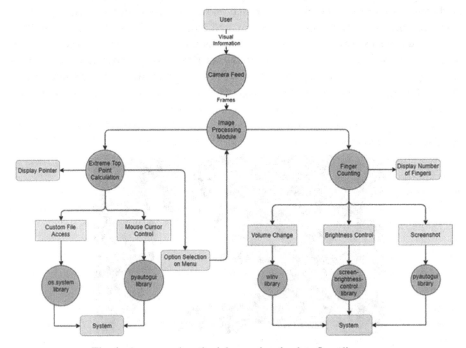

Fig. 4. A proposed methodology using the data flow diagram

will be needed to be set up to connect the machines for the transfer of data. The software proposed in the paper can run easily without the extensive use of specialized tools. Thus, it has very far-reaching accessibility and compatibility factor. A Python environment is needed on the system itself or on a central server for the processing, along with external libraries such as OpenCV and iMutils. The implementation requires a plain background with appropriate light; otherwise, it may give inaccurate results of where the finger is pointing or the number of fingers. The process of the proposed technique is depicted in Fig. 4.

3.2 Proposed Modules

The proposed technique includes some of the modules explained below.

- **Feed Capture and Background Averaging.** Capture the input from the attached webcam and calculate the background using OpenCV functions. Also, flip the image using iMutils.
- **Thresholding and Contour Drawing.** Binarize the frame with hand using thresholding and then draw contour along hand outline.
- **Segmentation and Convex Hull.** Separate mask of hand using AND operation, which compares every pair of pixels in the frames. Thus, the hand is separated as it is the only difference in every frame, compared to the background. Draw a circle using extreme points of the segment and calculate intersections with fingers.
- **System Utility Controls.** Track a point on hand segment for cursor control and custom file selection. Utilize finger count for volume control, brightness control and screenshot functionality.

4 Experimental Analysis and Results

While being based on the algorithm proposed by Asnaterabi Malima et al., the paper also incorporated a few heuristic changes made by the authors for an increase in precision. The threshold value used for separation of hand and background is 25, which seemed to work best with a plain well-lit background. Additionally, the height calculated from the algorithm was lowered by around 20 pixels, to improve the intersection with the little finger and thumb and increase accuracy. Figure 5 illustrates the output of the implemented algorithms before and after with respect to height.

The accuracy of finger detection was also dependent on the hand's angle concerning the camera. Also, shadows of the hand casted inside the ROI during detection were affecting the accuracy. Hence, proper lighting is also paramount for successful detection. However, it can be easily controlled during implementation. A menu was provided using a region-of-interest (ROI), via which the user can select any of the five functionalities presented. Another ROI was defined for the actual input anchored to the selected functionality. These 5 functionalities display the flexibility of approach that can be applied on the basis of visual input.

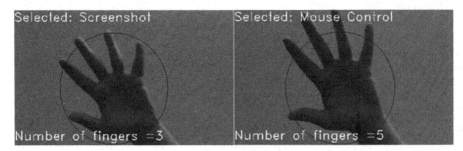

Fig. 5. Comparison of algorithm output before and after alteration in height

4.1 Functionalities of the Proposed Methodology

The following functionalities are imposed in the proposed methodology: volume control, brightness control of the device, cursor control, taking screenshots, and custom file access.

1. **Volume Control**
   The volume was accurately changed according to the number of fingers. As human-beings can normally have 5 fingers, the volume control level varies from 1 to 5. Also, to increase the volume level, the multiples of 5 can be programmed.

2. **Brightness Control**
   Brightness was changed accurately but faced the same issue as presented in the volume control.

3. **Mouse Cursor Control**
   The Region of Interest (ROI) that was used was small. It was scaled to full screen. Thus resulted in large mouse cursor movement for little hand movement, which led to slightly poor accuracy. It can be rectified using a more considerable ROI. The right and left click functionalities worked accurately from hand gestures.

4. **Screenshot**
   Screenshots were taken accurately via a reading of the hand gesture specified in the software. The function can be expanded to save the screenshots in a specified folder.

5. **Custom File Access**
   Three different file types were specified utilizing system path provided to the script: an executable file, a pdf file, and an image file. All could be successfully opened using their system calls by specifying their paths.

The RAM usage for each functionality over 10 s, with continuous usage, was calculated. It was found out that the software is not memory intensive and can run on low-end machines.

The average memory use over 10 s was also calculated for each functionality and was found to be highest for screenshot and lowest for volume control. The memory use was well within 150 MB. Figure 6 gives much detailed information about the memory usage over 10 s per every functionality. Table 1, lists the five functionalities versus the memory usage in GB. Figure 7 gives clear idea that the screen shot functionality takes much GB than other functionalities.

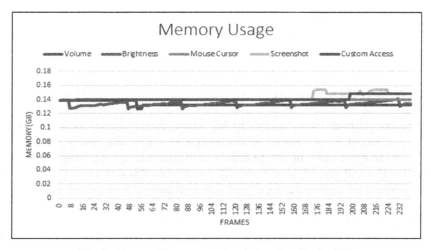

Fig. 6. Memory Usage over 10 s per functionality (in GB)

Table 1. Average memory usage per functionality (in GB)

Volume	Brightness	Mouse cursor	Screenshot	Custom access
0.133392445	0.133920606	0.139837154	0.142526293	0.14110899

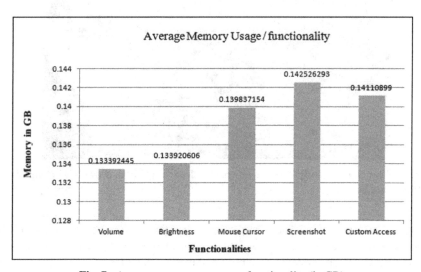

Fig. 7. Average memory usage per functionality (in GB)

The highest memory usage was found during continuous use of screenshot functionality over a 10 s period. The memory usage was within 150 MB. The above analysis provides clear idea about the proposed method's performance that it can be easily run on even lower end machines and can be integrated into the functioning of public devices that are not built with high performance capabilities.

5   Results and Discussion

The experimental results were obtained by the control of various functions using a laptop. Currently, the hand position and angle affect the program's accuracy and the background used. The paper presents a prototype demonstrating the use of basic image processing for hand gestures. Regarding performance, the proposed method was able to run on even older machines with reduced capacity and still provided real-time output. The memory capacity was found to be around 140 MB on average for the various functionalities.

Given that modern systems have more than 4 GB of memory, usually can run easily with the proposed method alongside other system utilities. Thus, the system can keep running in the background to be used whenever the user requires it. Figure 8 and Figure 9 give the utilization of ROI with a hand segmentation mask to show the increase in the volume control using the red point on the screen.

Fig. 8. Utilization of Menu ROI, displaying point selection (red dot), shown here alongside the hand segmentation mask

Figure 10 depicts the custom file selection process by using the red spot in the particular menu. The script is lightweight and does not consume system resources on too large a scale. Public systems use a proposed method that runs on low-end computers built only for the scope of specific, repetitive use. The algorithms and the UI may be dependent on the end requirement of the user.

Fig. 9. Selection of volume option from Menu ROI and displaying 1 finger in the segmentation ROI results in volume changing to 20%

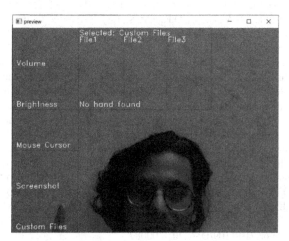

Fig. 10. Segmentation ROI converted into UI for selection of custom files when Custom File functionality selected from Menu ROI

6   Conclusion and Future Enhancement

The paper shows that it is possible to implement such a touchless system in common public machinery to reduce the risk of spreading the virus through common touch. A small prototype of the said implementations of using cameras to take visual input for system controls has been implanted. It is limited to a number of fingers currently and pointing using fingers. However, it can be extended to other gestures and the functions it is used for. The research leads to the conclusion that the paper has the potential also to control other basic features of the system, and can be used as a fully-fledged utility.

As computer vision algorithms get more accurate, the paper's research can be made even more effective and implementation-ready, as accuracy can be a significant hurdle in utilizing the proposed system under different scenarios. Reducing the dependency of the system on proper background and lighting is also something that can be improved. Implementing such a system in devices other than personal systems is a major area of focus. Developing the proposed method in a cross-platform architecture would lead to the integration of multiple services under a single bundle, which would lead to better flexibility and, thus, more adoption in daily lives. The proposed methodology has the means to provide accessibility support to deaf and dumb people, who cannot benefit from speech to text-based system-controls.

References

1. Malima, A., Ozgur, E., Cetin, M.: A fast algorithm for vision-based hand gesture recognition for robot control. In: 2006 14th IEEE Signal Processing and Communications Applications, Antalya, pp. 1–4 (2006). https://doi.org/10.1109/SIU.2006.1659822
2. Karkera, K., Thakar, J., Velani, R., Solanki, C., Mhatre, Y.: Vision based real time gesture recognition. In: SSRN Electronic Journal (2019). https://doi.org/10.2139/ssrn.3372082
3. Mukherjee, S., Ahmed, S.A., Dogra, D.P., Kar, S., Roy, P.P.: Fingertip detection and tracking for recognition of air-writing in videos. Expert Syst. Appl. **136**, 217–229 (2019). https://doi.org/10.1016/j.eswa.2019.06.034
4. Biswas, A.: Finger detection for hand gesture recognition using circular hough transform. In: Bera, R., Sarkar, S.K., Chakraborty, S. (eds.) Advances in Communication, Devices and Networking. LNEE, vol. 462, pp. 651–660. Springer, Singapore (2018). https://doi.org/10.1007/978-981-10-7901-6_71
5. Tang, H., Xiao, W., Liu, H., Sebe, N.: Fast and robust dynamic hand gesture recognition via key frames extraction and feature fusion. Neurocomputing **331**, 24–433 (2018). https://doi.org/10.1016/j.neucom.2018.11.038
6. Harini, V., Prahelika, V., Sneka, I., Adlene Ebenezer, P.: Hand gesture recognition using openCv and python. In: Smys, S., Iliyasu, A.M., Bestak, R., Shi, F. (eds.) ICCVBIC 2018, pp. 1711–1719. Springer, Cham (2020). https://doi.org/10.1007/978-3-030-41862-5_174
7. Xing, K., et al.: Hand gesture recognition based on deep learning method. In: IEEE 3rd International Conference on Data Science in Cyberspace (DSC), Guangzhou, pp. 542–546 (2018) https://doi.org/10.1109/DSC.2018.00087
8. Kane, L., Khanna, P.: Depth matrix and adaptive Bayes classifier based dynamic hand gesture recognition. Pattern Recogn. Lett. **120**, 24–30 (2019). https://doi.org/10.1016/j.patrec.2019.01.003
9. Shanthakumar, V.A., Peng, C., Hansberger, J., Cao, L., Meacham, S., Blakely, V.: Design and evaluation of a hand gesture recognition approach for real-time interactions. Multi. Tools Appl. **79**(25–26), 17707–17730 (2020). https://doi.org/10.1007/s11042-019-08520-1
10. Purushothaman, A., Palaniswamy, S.: Development of smart home using gesture recognition for elderly and disabled. J. Comput. Theor. Nanosci. **17**, 171–181 (2020). https://doi.org/10.1166/jctn.2020.8647
11. Benitez-Garcia, G., et al.: Improving real-time hand gesture recognition with semantic segmentation vol 21. Sensors **21**(2), 356 (2021). https://doi.org/10.3390/s21020356
12. Santoshi, G., Parwekar, P., Gowri Pushpa, G., Kranthi, T.: Multiple hand gestures for cursor movement using convolution neural networks. In: Satapathy, S.C., Bhateja, V., Janakiramaiah, B., Chen, Y.-W. (eds.) Intelligent System Design. AISC, vol. 1171, pp. 813–825. Springer, Singapore (2021). https://doi.org/10.1007/978-981-15-5400-1_77

13. Daniel, M.S., John, N.P., Prathibha Devkar, R., Abraham, R., George, R.E.: Speaking mouth system for dumb people using hand gestures. In: Suresh, P., Saravanakumar, U., Hussein Al Salameh, M. (eds) Advances in Smart System Technologies. Advances in Intelligent Systems and Computing, vol. 1163. Springer, Singapore, pp. 613–621 (2021). https://doi.org/10.1007/978-981-15-5029-4_51

14. Runwal, R., et al.: Hand gesture control of computer features. In: Kalamkar, V.R., Monkova, K. (eds.) Advances in Mechanical Engineering. LNME, pp. 799–805. Springer, Singapore (2021). https://doi.org/10.1007/978-981-15-3639-7_96

15. Kumari, S., Mathesul, S., Shrivastav, P., Rambhad, A.: Hand gesture-based recognition for interactive human computer using tenser-flow. Int. J. Adv. Sci. Technol. **29**(7), 14186–14197 (2020). https://doi.org/10.13140/RG.2.2.13563.95527

16. Ahmed, M., Hossain, M.A., Abadin, A.F.M.: Implementation and performance analysis of different hand gesture recognition methods **19**, 13–19 (2019). https://doi.org/10.34257/GJC STDVOL19IS3PG13

17. Zhang, L., Zhang, Y., Niu, L., Zhao, Z., Han, X.: HMM static hand gesture recognition based on combination of shape features and wavelet texture features. In: Jia, M., Guo, Q., Meng, W. (eds.) WiSATS 2019. LNICSSITE, vol. 281, pp. 187–197. Springer, Cham (2019). https://doi.org/10.1007/978-3-030-19156-6_18

18. Nguyen, D-L., Putro, M.D., Jo, K-H.: Hand Detector based on efficient and lightweight convolutional neural network. In: 2020 20th International Conference on Control, Automation and Systems (ICCAS), Busan, Korea (South), pp 2642–3901 (2020). https://doi.org/10.23919/ICCAS50221.2020.9268320

19. Anitha, A.: Home security system using internet of things. In: IOP Conference Series: Materials Science and Engineering, vol. 263, no. 4, p. 042026, IOP Publishing (2017). https://doi.org/10.1088/1757-899X/263/4/042026

20. Nowosielski, A., Chodyła, Ł.: Touchless input interface for disabled. In: Burduk, R., Jackowski, K., Kurzynski, M., Wozniak, M., Zolnierek, A. (eds.) Proceedings of the 8th International Conference on Computer Recognition Systems CORES, Advances in Intelligent Systems and Computing, vol 226. Springer, Heidelberg, pp 701–709 (2013). https://doi.org/10.1007/978-3-319-00969-8_69

21. Suleiman, A., Sharef, Z.T., Faraj, K., Ahmed, Z., Malallah, F.: Real-time numerical 0–5 counting based on hand-finger gestures recognition. J. Theor. Appl. Inf. Technol. **95**, 3105–3115 (2017)

22. Chanu, O.R., Pillai, A., Sinha, S., Das, P.: Comparative study for vision based and data based hand gesture recognition technique. In: 2017 International Conference on Intelligent Communication and Computational Techniques (ICCT), Jaipur, pp. 26–31, (2017). https://doi.org/10.1109/INTELCCT.2017.8324015

23. Foroutan, Y., Kalhor, A., Nejati, S.M., Sheikhaei, S.: Control of computer pointer using hand gesture recognition in motion pictures. In: Computer Vision and Pattern Recognition, Human-Computer Interaction (cs.HC) (2020). https:/arXiv:2012.13188

24. Pisharady, P., Vadakkepat, P., Loh, A.P.: Attention based detection and recognition of hand postures against complex backgrounds. Int. J. Comput. Vision **101**, 403–419 (2013). https://doi.org/10.1007/s11263-012-0560-5

25. Priyanka, R., Prahanya, S., Jayasree, L.N., Angelin, G.: Shape-based features for optimized hand gesture recognition. Int. J. Artif. Intell. Mach. Learn. (IJAIML) **11**(1), 23–38 (2021). https://doi.org/10.4018/IJAIML.2021010103

Virtually Interactive User Manual for Command and Control Systems Using Rule-Based Chatbot

Shruti Jain$^{(\boxtimes)}$, Shivani Kapur, and Vipin Chandra Dobhal

Central Research Laboratory, Bharat Electronics Limited, Ghaziabad, Uttar Pradesh, India
{shrutijain94,shivanikapur,vipinchandradobhal}@bel.co.in

Abstract. Command and Control (C2) Systems are complex information systems consisting of humans, integrated hardware, and course of action. Traditionally, a documented user manual is provided to the operators to facilitate them with a strong understanding of the system. However, with increasing number of complex functionalities, it becomes difficult and time-consuming for operators to comprehend the document. This study proposes a conceptual framework of Interactive User Manual (C2IUM), which uses a rule-based chatbot and provides clear instructions in a chat-like interface using Natural Language Processing. The paper outlines three-layered methodology including (i) Dataset creation, (ii) Model building, and (iii) Integration. To verify the applicability of the tool, an experiment has been performed on C2 surveillance system and an accuracy of 82% is obtained. The proposed tool introduces automation and enables better customer support in terms of 24 * 7 accessibility, swift answers, and well-structured responses. The tool is robust and scalable.

Keywords: Natural Language Processing · Deep learning · C2 system · C4I system · Command and Control system · Interactive user manual · Rule-based chatbot

1 Introduction

Command and Control (C2) systems and Command, Control, Communications, Computer and Intelligence (C4I) systems are distributed supervisory control systems [1]. They have applications in monitoring and planning, smart city development, military and defence, air traffic control, railway signalling, network-centric warfare, mission control etc. They are complex systems consisting of numerous components. These components can themselves be complicated and can interact with and affect other systems. These systems are designed to manage and process information, react to the changes in any situation, reduce the chaos, and provide commanders with the ability to interact with staff and available forces and take necessary action accordingly [2]. Since the system is monitored in a dynamic environment, the system's complications are increased manifold if the operator/user does not have proper understanding of the system. Improper usage and little knowledge of the system can cause serious issues. The relationship between command and control and also within the components of C2, needs to be well known. Well trained staff is essentially required to fight the uncertainty.

© Springer Nature Switzerland AG 2021
A. Solanki et al. (Eds.): AIS2C2 2021, CCIS 1434, pp. 162–172, 2021.
https://doi.org/10.1007/978-3-030-82322-1_12

Companies provide well-written documents in the form of instruction manuals that can help users understand the system. Manuals give consistency and aids in avoiding information gap between application developers and users. Manuals are of various types, including product manual, installation manual, troubleshoot manual, user manual, operations manual, etc. [3]. However, Neil Fiore, a Berkeley, Calif., psychologist, once said, "Manuals just slow you down and make you feel stupid. The directions are too slow, too detailed and use too much abstract, arcane or academic language" [4]. In a fast-changing environment, people find reading manuals as a time-consuming and futile activity. Also, if instructions in the manual are not clear, operators wait for the next available operator to clarify their doubt.

With recent advancements in Artificial Intelligence, chatbots are already handling user queries in various domains by automating the conversation. According to [5], "The chatbot market size is projected to grow from 2.6 billion in 2019 to 9.4 billion by 2024 at a compound annual growth rate (CAGR) of 29.7%".

This paper proposes Interactive User Manual (IUM) for Command and Control System (C2). The proposed tool, namely C2IUM, shall overcome the limitations of existing technology (Sect. 2.1). C2IUM shall provide information to the user regarding the functions of the system using a rule-based chatbot. The operator shall enter the query and receive a response, thereby gathering deep knowledge of the system swiftly and systematically. Section 2 (Literature Review) of the paper describes the proposed solution's background and the related work. A three layered architecture of the proposed design is outlined in Sect. 3. Section 4 validates the applicability of the tool in C2 systems using an experiment and Sect. 5 evaluates the features of the tool based on the design of Sect. 3. Finally, in Sect. 6, conclusions are drawn and future work is discussed.

2 Literature Review

A chatbot, also known as a chatterbot, is an Artificial Intelligence-based tool that can be integrated with multiple applications and allows customer-centric conversations by imitating human discussions. [6] sheds light on why people prefer to use chatbots. The paper suggests designing guidelines reflecting the motivation of the user. An in-depth survey on different types of chatbots based on their functionalities (rule-based, AI-based and knowledge-based) is conducted and presented in [7]. Authors in [8] have explored various open-source frameworks for building an enterprise chatbot.

Recent advancements are done in integrating the "Internet of Battlefield Things" and C3I systems and their impact on military decision making. It can be observed that Machine Learning technology is introducing a new era of automation in these systems [9]. A detailed literature review laying down the foundation of the network in C2 systems using network centric warfare theory is presented in [10].

According to the empirical study in [11], the authors have concluded that software users prefer information sources that take little effort. Authors argue that users look forward to use documentation that is well designed and structured. Documented user manuals are usually present with Command and Control systems, but little work is done on introducing automated interactive manuals within the system. Interactive product manual for home appliance systems has been designed and implemented using chatbots in [12]. The aid of an experiment also justifies usefulness of the tool. Similarly, using an AI-based tool, authors in [13] have used frequently asked questions (FAQs) as source dataset and trained a bot for providing responses to the business queries. Authors in [14] have integrated a chatbot with a business intelligence tool. Further, Machine Learning approach is also used in the finance and healthcare industries to develop a chatbot and test its applicability [15, 16].

2.1 Existing Technology

Instructions in products and applications are provided in graphical videos, small animations, product manuals or audio repositories. However, in Command and Control systems, users' guide is usually present as a documented text. Web-based C2 applications are stored as Hypertext Markup Language (HTML) pages and for desktop-based applications; a Portable Document Format (PDF) is used. Some applications use Compiled HTML (CHM) files for the same. CHM files are Microsoft proprietary help files that consist of compiled HTML pages and navigation tools.

Documented texts are long and sometimes difficult to read. To search for an application's feature in PDF, the user must enter the exact keyword without any mistake. Most of the time, there are multiple matches to the keyword and the user has to go through all of them to find the one they are looking for, hence increasing the time taken. Similarly, for HTML based user manuals, searching on one page limits the search database to that particular page. Upcoming sections propose a rule-based chatbot as a user manual to overcome these limitations.

3 Proposed System

This paper aims to develop an automated and interactive user manual that can be integrated with C2 and C4I systems. This section gives a detailed explanation of the proposed methodology. Figure 1 illustrates the system overview.

The operator shall login into the system. In the case of authentication failure, the operator shall exit the system. In case of a successful login, operator can select the Interactive User Manual option from the system display and enter the query. Based on user query, a suitable response shall be generated. The operator can exit the manual when all queries are resolved.

3.1 An Overview

Chatbots can broadly be classified as rule-based and self-learning based [17]. In rule-based chatbots, responses to the query are fixed and hard-coded in the dataset. Bot answers the question based on the rules on which it is trained and does not learn from the conversations. On the other hand, self-learning chatbots use AI to generate the response. They involve the understanding of user query and formulating a dynamic response. They are suitable in case when the bot requires decision-making.

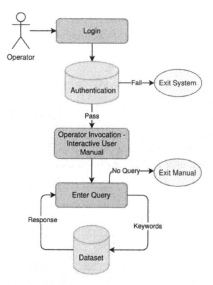

Fig. 1. System overview.

To obtain any goal or complete any procedure, the set of rules in C2 systems is well-patterned and fixed. That is, the steps to perform functionality do not change. For example, to read any sensor data, the operator can select the corresponding sensor from the operator display and examine raw sensor videos, graphics and other sensor-related information. Similarly, to add a new sensor, steps are predefined [2]. Since there is a series of predefined and highly structured rules, deploying a rule-based chatbot is the best choice.

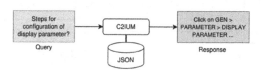

Fig. 2. C2IUM: Input and output

C2IUM shall take the user query as input, process it to find out the intent and finally output the response, as shown in Fig. 2.

3.2 Methodology

To perform the experiments, Google Colaboratory (also known as Colab) is used. All implementations are done using Python 3 Google Compute Engine Backend (GPU) settings with 12.72 GB random access memory and 358.27 GB disk space over the cloud. Text analysis is done using Natural Language Processing and Natural Language Toolkit (NLTK) library of Python is used to perform the pre-processing of the data

[18]. It is widely used for research and prototyping. The following section explains the proposed methodology and details of the implementation.

Three Layered Methodology
To implement the proposed system, a three-layered methodology, as shown in Fig. 3, is suggested. It includes

- Generating the dataset
- Processing the data and model building
- Integrating the developed user manual with the C2 system

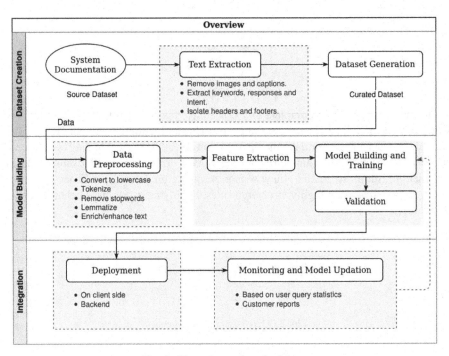

Fig. 3. Three-layered methodology

As discussed in Sect. 2.1, PDF file or HTML pages are used as information manuals and contain system documentation. For the proposed rule-based chatbot C2IUM, the dataset shall be generated using the existing technology; that is, the data from documentation shall be converted to a well-formatted JavaScript Object Notation (JSON) file. A Java-based conversion tool is developed for the same, which shall take documentation as input and produce a curated JSON file as output by extracting the useful information. The conversion tool shall remove images and their captions used in documentation files. Further, headers and footers shall be removed and non-relevant information shall be skipped. Generated JSON file shall specify the intents, keywords and responses corresponding to all the functionalities of the C2 system. Alternatively, an eXtensible mark-up Language (XML) file can be created specifying the same details. A sample JSON file consisting of two functions is shown in the greybox below.

```
{"Options":
    [
        {
        "intent" : "version number",
        "keywords" : ["What is the current version number", "Version details of system"],
        "response" : ["Version No. ABC-1.0"],
        },

        {
        "intent" : "configure display parameter",
        "keywords" : ["How to configure display parameters", "Enable operator to
        configure parameters for various elements of tactical picture",
        "steps for display configuration"],
        "response" : ["1. Click on 'GEN>PARAMETER>DISPLAY PARAMETER'.
        2. Select 'Change Parameters'. For more details *Link here*. "],
        }
    ]
}
```

"Intent" specifies the intention of the query. The number of intents is equal to the number of classes in which classification is to be done. "Keywords" specify the words that might be used by the operator while entering the query. "Response" is the output corresponding to intent.

Layer Two: Processing/Model Building
Firstly, the JSON file obtained in the data layer is pre-processed. The complete text is divided into tokens using the NLTK library of Python. Other steps of pre-processing include conversion to lowercase characters, lemmatization and removal of stopwords and duplicates. Modified text is then made machine-understandable by converting it into numbers using Bag of Words.

The next step involves building the neural network for which Keras [19] sequential API is used. The architecture of the neural network is shown in Fig. 4. The input layer consists of a one-hot encoded vector whose size is equal to the length of a bag of words. Input is sent to 2 fully connected hidden layers, each consisting of 128 and 64 neurons. Finally, the output layer is a softmax layer with a number of neurons equal to several possible intents. The probability of query indicating intent, is given as output. The intent with highest probability is selected as a response to the query. A threshold value of probability is set depending on the dataset. If none of the intents qualifies above the threshold, then it is safe to say that the user's query does not match the dataset's keywords. In such a situation, a report indicating the query is generated and stored in a separate database. This report can be used at the time of system monitoring and updating.

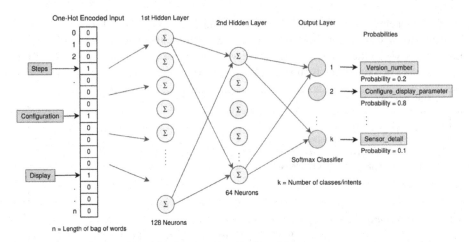

Fig. 4. Neural network architecture

It is to be observed that the displayed architecture in Fig. 4 is a sample network and it can be curated according to the need of the dataset.

Layer Three: Integration
In this final layer, the model shall be deployed on the client-side. Screenshots of the deployed Interactive User Manual are shown in Fig. 5. It can be observed that the chatbot responds to different aliases of the word "number" in the same way. Another salient feature is that the bot is immune to spelling mistakes. For example, the bot understands "parametr configraton" as "parameter configuration" and responds accordingly. Moreover, the bot is not case sensitive. The color of the display shall be configurable and can be changed by the user.

It's important to note that all user query statistics and other reports shall be stored in the database. These statistics shall be utilized for system monitoring and model updation. In case a user enters a query and doesn't find a suitable response, a report stating the same can be generated and stored in the database. Based on the reports, reinforcement can be done and updations can be made in the JSON dataset. Similarly, if required, the model can be built and retrained, followed by validation and deployment. This iterative methodology provides high flexibility, easy debugging and faster enhancements.

It can be noted that the tool is generic; that is, based on the content of JSON or XML file, the tool can be integrated with different domains of command and control systems.

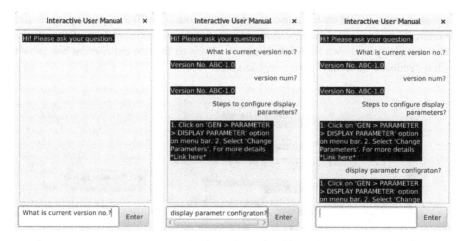

Fig. 5. Screenshots of interactive user manual

4 Experiment

To validate the applicability of the proposed tool, a case study based approach was deployed. Documented manual of "Parameter Surveillance C2 system" consisting of 189 functions was used as a source dataset. Dataset consists of various features like serial number, page number, name of the function, objective, steps of performing action, expected result, and exception conditions in a PDF file. Irrelevant features like serial and page number were dropped and appropriate data was extracted using NLP. This data was then fed to the developed Java based conversion tool, as explained in Sect. 3. After text extraction and conversion, a curated dataset was obtained in the form of a JSON file. The data was pre-processed and the model was built as per neural network architecture in Fig. 4. The model was trained for 200 epochs and finally, the tool was integrated with the system. The results on this dataset show an accuracy of 82%.

Further, an experiment was performed on 10 participants (6 males and 4 females) on 11 November 2020. Each participant was asked to operate the parameter surveillance system. It was made sure that none of them had any prior knowledge of the system. Participants were provided with both traditionally documented as well as the developed interactive manuals (C2IUM). Their task was to perform 7 common functionalities on the system in 15 min by referencing the manuals. At the end of the experiment, participants were asked to fill a questionnaire regarding their experience. Observations were noted based on the following parameters: (A) Accuracy of performing functionalities, (B) Effectiveness, (C) Ease of use, (D) Less time consumption and (E) Manual Preference.

5 Results and Discussion

Observations from the experiment show that 80% of participants preferred to use interactive manual, whereas only 20% were comfortable using documented guides. 60% of participants said that they got accurate results using C2IUM (Interactive Manual). 90%

of people found C2IUM easier to use. It can also be noted that 70% of participants found the interactive manual less time-consuming. A graphical representation of the results of the case study is shown in Fig. 6. It can be observed that the experiment has shown positive feedback in favour of the proposed tool.

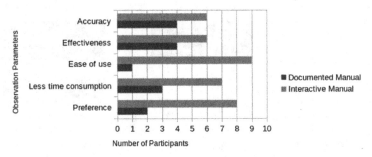

Fig. 6. Observation results

Features of C2IUM can be classified into two parts: benefits for the user or operator of command and control systems and benefits from the business point of view. Table 1 outlines the features of the tool.

Table 1. Features of C2IUM

For user	For business
• Swift answers to simple questions • Instant solutions at any time of the day • Direct links to query related detailed explanation • Immunity to spelling mistakes in user typed query	• They enable ground level executives, operators, admins, and operational managers to better understand various functionalities supported by the system • Competition with the business industry by introducing automation through chatbot • Convenience and ease of use to the customer • Opportunity for companies to collect and analyze user's data to have better understanding of requirements and provide better and personalized service

However, like any other tool, the proposed tool is subjected to few limitations. Since an NLP engine's training is not visible to human eyes, it becomes complex to understand, train, and maintain the engine. The proposed approach depends on a pre-existing source dataset (system documentation) to create the curated dataset.

Any anomaly in documentation shall be directly reflected in the generated dataset. Engine might not identify the user's query's intent with 100% accuracy and the response might suffer in a few cases.

6 Conclusion and Future Work

This paper introduced a conceptual framework of an NLP based tool named C2IUM. Firstly, the need for having an automated user manual in command and control systems is described. The proposed tool takes user query as input and provides a rule-based response in a chat-like interface. Next, a three-layered methodology has been deployed to introduce modularity in the process. It has proved to be effective and significant. Further, the tool's role is discussed from the user's point of view and the perspective of the business. It can be argued that the proposed tool is more interactive and user-friendly than the traditional technology of documented manuals.

In future studies, work can be done to minimize the limitations of the system. Data can be crowd-sourced for generic requirements. Additionally, other features like voice-based assistance and query sentence completion can be introduced.

References

1. Shattuck, L.G., Woods, D.D.: Communication of intent in military command and control systems. In: The Human in Command, pp. 279–291, Springer (2000). https://doi.org/10.1007/978-1-4615-4229-2_19
2. FAS United States Navy, chapter 20 command, control, and communication. https://fas.org/man/dod-101/navy/docs/fun/part20.htm. Accessed 11 Oct 2020
3. Manual of operations: key for knowledge management. https://foundersguide.com/manualof-operations-helpful-tool-for-business/. Accessed 11 Oct 2020
4. Mayer, C.: Why Won't we Read the Manual, p. H01, Washington Post (2002)
5. The latest market research, trends, and landscape in the growing ai chatbot industry. https://www.businessinsider.com/chatbot-market-stats-trends?IR=T. Accessed 14 Oct 2020
6. Brandtzaeg, P.B., Følstad, A.: Why people use chatbots. In: Kompatsiaris, I., et al. (eds.) INSCI 2017. LNCS, vol. 10673, pp. 377–392. Springer, Cham (2017). https://doi.org/10.1007/978-3-319-70284-1_30
7. Nuruzzaman, M., Hussain, O.K.: A survey on chatbot implementation in customer service industry through deep neural network. In: 2018 IEEE 15th International Conference on e-Business Engineering (ICEBE), pp. 54–61. IEEE (2018)
8. Singh, A., Ramasubramanian, K., Shivam, S.: Building an enterprise chatbot: work with protected enterprise data using open source frameworks. Apress (2019)
9. Russell, S. Abdelzaher, T.: The internet of battlefield things: the next generation of command, control, communications and intelligence (c3i) decision-making. In: MILCOM 2018–2018 IEEE Military Communications Conference (MILCOM), pp. 737–742. IEEE (2018)
10. Eisenberg, D.A., Alderson, D.L., Kitsak, M., Ganin, A., Linkov, I.: Network foundation for command and control (c2) systems: literature review. IEEE Access 6, 68782–68794 (2018)
11. van Loggem, B.: 'Nobody reads the documentation': true or not?. In: Proceedings of ISIC: The Information Behaviour Conference, no. Part 1 (2014)
12. Choi, H., Hamanaka, T., Matsui, K.: Design and implementation of interactive product manual system using chatbot and sensed data. In: 2017 IEEE 6th Global Conference on Consumer Electronics (GCCE), pp. 1–5. IEEE (2017)
13. Thomas, N.: An e-business chatbot using AIML and LSA. In: 2016 International Conference on Advances in Computing, Communications and Informatics (ICACCI), pp. 2740–2742. IEEE (2016)

14. Vashisht, V., Dharia, P.: Integrating chatbot application with Qlik sense business intelligence (BI) tool using natural language processing (NLP). In: Micro-Electronics and Telecommunication Engineering, pp. 683–692. Springer (2020)
15. Okuda, T., Shoda, S.: AI-based chatbot service for financial industry. Fujitsu Sci. Tech. J. **54**(2), 4–8 (2018)
16. Nadarzynski, T., Miles, O., Cowie, A., Ridge, D.: Acceptability of artificial intelligence (AI)-led chatbot services in healthcare: a mixed-methods study. Digit. Health **5**, 2055207619871808 (2019)
17. Build your own smart ai chat bot using python. https://medium.com/@randerson112358/build-your-own-ai-chat-bot-using-python-machinelearning-682ddd8acc29. Accessed 16 Oct 2020
18. Bird, S., Klein, E., Loper, E.: Natural Language Processing with Python: Analyzing Text with the Natural Language Toolkit. O'Reilly Media, Inc. (2009)
19. Géron, A.: Hands-On Machine Learning with Scikit-Learn, Keras, and TensorFlow: Concepts, Tools, And Techniques To Build Intelligent Systems. O'Reilly Media (2019)

Healthcare in Smart Cities

Wrapper-Based Best Feature Selection Approach for Lung Cancer Detection

Vidhi Bishnoi⬥, Nidhi Goel$^{(\boxtimes)}$⬥, and Akash Tayal⬥

Indira Gandhi Delhi Technical University for Women, New Delhi, India

Abstract. Lung cancer, the leading cause of death due to cancer all over the world. Proper diagnostic system can help the radiologists in early diagnosis of cancer. Computer based systems provides the ease to detect the lung cancer efficiently. Feature selection plays important role in the performance of such automated models. The wrapper based Hybrid Sequential Exhaustive Feature Selection (HSEFS) method has been proposed to find the optimal solution for classification. The proposed method has been compared with three algorithms, sequential forward selection (SFS), sequential backward selection (SBS), and exhaustive feature selection (EFS) on 300 lung CT scan nodules to evaluate best feature score. Random Forest (RF) classifier has been used in association with these wrapper methods. The cross-validation score 0.99 was achieved by HSEFS method which serve the best feature selection method among all the above mentioned methods.

Keywords: Hybrid feature selection · Lung cancer · Sequential forward selection · Sequential backward selection · Exhaustive feature selection

1 Introduction

Lung cancer cases are large in count amongst other cancer cases [1]. It is estimated that lung cancer is the lead factor of death world – wide [2]. The majority of cases are often caused due to continuous smoking which gives the lowest survival rate of the lung cancer of all cancers [3]. Therefore, proper diagnosis of lung cancer is very important at its initial stages to save the lives. Lung cancer images are obtained from chest radio scans and computed tomography (CT) scans and the diagnosis is finally confirmed by the radiologist after the analysis of the scans. But, the proper examination and diagnosis of the lung CT image in the biomedical field is quite sensitive and time consuming process [4]. These reasons motivate the researchers to utilize the new computing technologies with computer-based automated systems to detect and classify the lung cancer. Thereby, reduces the cost and time of diagnosis.

For this purpose, the automated techniques (i.e. machine learning algorithms) are used to perform the classification of lung cancer images in malignant and benign. The algorithms access the huge datasets with cheaper and powerful

© Springer Nature Switzerland AG 2021
A. Solanki et al. (Eds.): AIS2C2 2021, CCIS 1434, pp. 175–186, 2021.
https://doi.org/10.1007/978-3-030-82322-1_13

computational processing along with feasible data storage. In the context of lung cancer detection, the performance of the various machine learning techniques depends on the data used to process. This includes the extraction and selection of various kinds of features for the learning of the classifier. The feature selection is an important part in data mining for any machine learning algorithm. It aims to find representative features which reduces the over fitting, improves the accuracy, and simplifies the model.

In general there are three categories of feature selection methods, classified as: filter methods, wrapper methods and embedded methods [5]. Filter methods takes up the relevance of the features by looking their properties and sorts them according to the score. Wrapper methods finds the optimal feature subset by sorting the features on the basis of classifier performance. The third type, embedded methods are also used to optimize the classifier performance and are almost similar to the wrapper methods. These methods are different from wrapper methods as an inherent model building metric is used during learning. The wrapper methods are efficient in solving real problems as in the case of biomedical processing since they are accurate and improves classification results. In this paper, a hybrid wrapper feature selection method is proposed to avoid the curse of dimensionality and to avoid over fitting of data in the classification process. For this proposed method HSEFS with sequential forward, sequential backward and, exhaustive feature selection methods have been implemented. The EFS is the most greedy search algorithm of all the wrapper methods since it explores all the combination of features and selects the best. But the proposed method HSEFS is found more efficient and consumes less time as compared to other methods.

The rest of this paper is organized as: Sect. 2 gives the related work on the lung cancer detection. Whereas, the proposed methodology is explained in Sect. 3. Section 4 elaborates the results and discussions and Sect. 5 reflects the conclusion.

2   Related Work

In this section the literature on the lung images feature selection and classification is presented. Lung cancer mentions to the abnormal growth of cells in any part of the lung; which tends to procreate in an uncontrolled way [6]. The cells are associated in the form of nodules, which can be seen in CT scan images. The classification process is carried out with the learning of features extracted from lung CT scan images. The selection of right features is a key to improve the classifier efficiency. Many feature selection methods have been used to select the best feature subset to feed the classifiers.

The principal component analysis has been applied by [7] to find the distinct features and built fuzzy kNN (FkNN) model for the diagnosis of Parkinson's disease. They achieved 96.07% of classification accuracy with this model. [8] applied a combined genetic-fuzzy algorithm on 32 patients, 56 features were selected to diagnose lung cancer and attained 97.5% accuracy. [9] employed two feature space reduction strategies: principal component analysis (PCA) and linear discriminant analysis (LDA) for the diagnostic of breast cancer. They found LDA better than PCA technique and achieved 97.06% and 95.88% accuracies respectively. [10] also applied LDA technique for feature selection. In their work, an optimal deep neural network (ODNN) was used to classify lung nodules as either malignant or benign. They reported accuracy of their algorithm were shown to be 94.56%.

Meta-heuristic algorithms such as multi-objective firefly, imperialist algorithm, genetic algorithm, and particle swarm optimization has been deployed by [11]. They determine the least number of features which can give the highest classification accuracy. They achieved best accuracy 95.12% from meta-heuristics MOFA method. [12] proposed Haralicks textures features of 3D images in a more realistic way. Area under the curve of receiver operating characteristic (ROC) curve was obtained for the extracted texture features. They achieved 97.17% and 89.1% sensitivity for 3D and 2D texture features respectively by an artificial neural network (ANN).

[13] proposed a embedded particle swarm feature selection method called HPSO-LS which uses a local search strategy to select the discriminated and salient feature subset. The Fisher method along with genetic optimization was proposed by [14] to address the common CT imaging signs of lung diseases (CISLs) recognition problem and obtained an optimal feature subset. The proposed approach is as follows.

3   Proposed Method

The proposed methodology may be the solution for the snag of machine learning algorithms, lists:

- The improvement in the classification accuracy and reducing the over fitting of the data.
- The reduction in the dimension of the feature set and selection of eccentric feature subset.
- Simplification of the machine learning model.

Despite all the challenges above, the present paper proposed the method shown in the flow chart Fig. 1 and tried to improve the diagnosis results of lung cancer. The lung CT scan image acquisition is done from the publically available dataset of Lung Image database consortium (LIDC). The dataset consists of 1018 cohorts which includes the multiple numbers of slices from clinical thoracic CT scan [15]. [16] makes the dataset accessible and reusable by generating a standardized Digital Imaging and, Communications in Medicine (DICOM) representation of the annotations results present in XML file. The radiologists marked the annotations in three groups: nodules ≥3 mm (lesion considered to be a nodule), nodules ≤3 mm (lesion considered as benign) and pulmonary lesions are non nodules of ≥3 mm and does not possesses features of nodules.

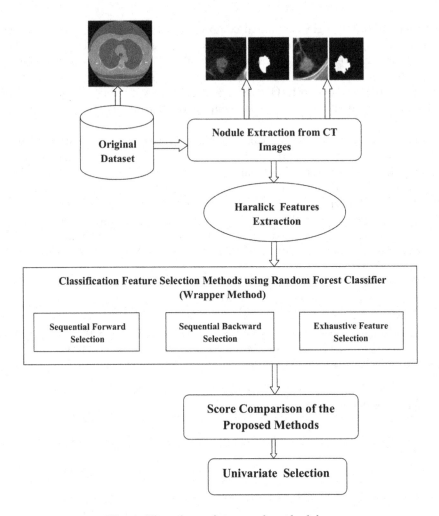

Fig. 1. Flow chart of proposed methodology

The nodules were extracted by the pylidc library package on python programming language described in (https://pylidc.github.io/) by [17]. This helped in acquiring the annotation contours of the individual scans and patients. The pylidc library provides an interface for the nodules and combines the annotations for every single scan by measuring distance between the annotations. It creates numpy arrays augmented to the dimensions of the CT scan image.

Further, in this paper 13 Haralick features were computed and applied to the HSEFS algorithm along with feature selection methods: sequential forward, sequential backward and, exhaustive feature selection. The proposed method

S.No	Haralick Features	Equation	Description
1	Angular Second Moment	$\sum_i \sum_j P(i,j)^2$	$P(i,j)$ is the $(i,j)^{th}$ element of the normalized GLCM
2	Contrast	$\sum_{n=0}^{N_g-1} n^2 \{ \sum_{i=1}^{N_g} \sum_{j=1}^{N_g} P(i,j) \}, |i-j| = n$	Where, Ng is the number of gray levels. P is the normalized symmetric (Gray Level Co-occurrence Matrix (GLCM) of dimension Ng × Ng.
3	Correlation	$\dfrac{\sum_i \sum_j (ij) P(i,j) - \mu_x \mu_y}{\sigma_x \sigma_y}$	Where, μ_x, μ_y and σ_x, σ_y are means & standard deviation. $\mu_x = \sum_i \sum_j i\, P(i,j), \ \mu_y = \sum_i \sum_j j\, P(i,j)$ $\sigma_x = \sqrt{\sum_i \sum_j (i-\mu_x)^2 P(i,j)}, \ \sigma_x = \sqrt{\sum_i \sum_j (i-\mu_x)^2 P(i,j)}$
4	Variance (Sum of Square)	$\sum_i \sum_j (i-\mu)^2 P(i,j)$	
5	Inverse- Difference Moment	$\sum_i \sum_j \dfrac{1}{1+(i-j)^2} P(i,j)$	
6	Sum -Average	$\sum_{k=2}^{2N_g} k P_{x+y}(k)$	Where, x and y are the row and column of the entry in the co-occurrence matrix, and $P_{x+y}(k)$ shows the probability of co-occurrence matrix coordinates.
7	Sum Variance	$\sum_{i=2}^{2N_g} (i-f_s)^2 P_{x+y}(i)$	
8	Sum Entropy	$-\sum_{k=2}^{2N_g} P_{x+y}(k) \log\{P_{x+y}(k)\} = f_s$	
9	Entropy	$-\sum_i \sum_j P(i,j) \log(P(i,j))$	
10	Difference -Variance	$\sum_{k=0}^{N_g-1} k^2 P_{x-y}(k)$	
11	Difference- Entropy	$-\sum_{k=0}^{N_g-1} P_{x-y}(k) \log\{P_{x-y}(k)\}$	
12	Informational Measure of Correlation 1	$\dfrac{HXY - HXY_0}{\max\{HX, HY\}}$	Where, $HXY = -\sum_i \sum_j P(i,j) \log(P(i,j))$ and HX and HY are the entropies of P_x & P_y. $HXY_0 = -\sum_i \sum_j P(i,j) \log\{P_x(i) P_y(j)\}$
13	Informational Measure of Correlation 2	$\sqrt{1 - \exp[-2(HXY_1 - HXY)]}$	$HXY_1 = -\sum_i \sum_j P_x(i) P_y(j) \log\{P_x(i) P_y(j)\}$

Fig. 2. Table for Haralick features with description

is known as Hybrid Sequential Exhaustive Feature Selection (HSEFS) method, used to select a discriminative feature subset. This section briefly elaborates extracted texture features and the proposed method.

3.1 Haralick Features

Haralick features contributes the texture features or global feature descriptor, counts the gray level co-occurrence matrix (GLCM). The matrix is of the dimension equal to the number of gray level of the region of interest (ROI) which is the extracted nodule in this paper. Figure 2 describes the various haralick features extracted for the learning of the classifier.

3.2   Sequential Forward Selection

This is a greedy search method for feature selection, finds optimal feature subset by iterating over the complete feature set. It reduces a N dimensional feature set to D dimension feature space, where D is less than N. Firstly, SFS removes all the features from the feature space, then adds one feature at a time in a new feature sub set while checking the performance of the classifier.

3.3   Sequential Backward Selection

It also belongs to the family of wrapper feature selection methods and it works in contrast of SFS. It removes the features one by one until the best classification accuracy criteria is achieved.

3.4   Exhaustive Feature Selection

Exhaustive feature selection method is the most greedy search algorithm to find the best features in small datasets. It selects best features with all the possible combinations of the complete set of features according to the classification score. But its computational time is quite large as compared to the above two algorithms. However, EFS is very efficient and consumes lesser evaluation time for small datasets.

3.5   Hybrid Sequential Exhaustive Feature Selection (HSEFS)

To solve this problem, proposed method has been applied to reduce the computational time and increase the accuracy. This section explains the proposed hybrid sequential exhaustive feature selection (HSEFS) method, which gives best results in biomedical area. The proposed algorithm resulted the best feature score among the above mentioned feature selection methods. The proposed algorithm is shown in Fig. 3. The feature set obtained by extracting the haralick features and an empty feature subset G is generated to store the best selected features. The complete data is divided into train and test data for the classification. Sequential part of the algorithm keeps adding the features into the subset G by initializing the number of features in subset to X_1 equal to zero. The features are added according to the test accuracy A_{c1} of the classifier at the first stage to achieve the target subset. These features were applied to the second part of the proposed method where, an new feature subset Y_n for n = 1,2,3,...,N is maintained to store the final best features. The random forest classifier has been used to evaluate the model accuracy and helps to update the best features in the combinations of the features subset. Finally, new best feature subset is achieved which results in good accuracy score.

Proposed HSEFS Algorithm pseudo code:

Input: Create a set of features F_i (i = 1, 2, 3, ..., k)
        Set the size of target feature subset N
Output: Best feature subset Y of size N.
    1. Create an empty set $G \leftarrow \Phi$
    2. Train data D_train and test data D_test
    3. Add features to G
       Initialize $X_i = 0$
       For F_i, $G \leftarrow F_i$ // add to feature subset
         $X_1 = X_1 + 1;$
       End
    4. Train the classifier by features in G and evaluate the test data accuracy A_{c1}.
    5. Go to the step 3.
    6. Select the feature subset G and train the classifier to get the accuracy A_{c2}.
    7. Make the possible combinations of the complete feature subset G and get a new best feature subset Y_n (n = 1, 2, ..., N).
    8. Update Y_n.

Fig. 3. Proposed algorithm

4 Result and Discussion

The feature selection methods have been evaluated based on univariate selection. The algorithms resulted in the best feature selection in context to the classifier performance which gave the score as shown in Fig. 7. The performance graph shows the performance score and the number of selected features. As a result 0.654, 0.658, 0.96 and 0.99 scores for SFS, SBS, EFS and, HSEFS have been achieved respectively. The best features selected by HSEFS were angular second moment, contrast, correlation and sum of squares.

Table 1. Comparison of the results achieved from SFS, SBS, EFS and, the proposed feature selection method

Methods	No. of selected features	Classifier	Score
SFS	7	RF	0.654
SBS	7	RF	0.658
EFS	7	RF	0.96
HSEFS (proposed method)	3–7	RF	0.99

Table 2. Accuracy scores obtained with proposed algorithm and others

Study	Method	Accuracy (%)
Chen et al. [7]	PCA	96.07
Alharbi and Abir [8]	Genetic Fuzzy	95.60
Joshi et al. [9]	LDA	97.06
Lakshmanaprabhu et al. [10]	LDA	94.56
Alirezaei et al. [11]	Meta-heuristic	95.12
Proposed method	HSEFS	99

feature_idx		cv_scores	avg_score	feature_names	ci_bound	std_dev	std_err
1	(5,)	[0.516666666666667, 0.6, 0.583333333333334,	0.583333	(Sum average,)	0.0661136	0.0424918	0.0245327
2	(5, 10)	[0.666666666666666, 0.633333333333333, 0.616..	0.629167	(Sum average, Difference entropy)	0.039514	0.0246503	0.0142319
3	(5, 10, 12)	[0.75, 0.633333333333333, 0.666666666666666, …	0.6625	(Sum average, Difference entropy, Info.measure…	0.0893601	0.0557462	0.0321851
4	(5, 8, 10, 12)	[0.683333333333333, 0.65, 0.633333333333333,…	0.654167	(Sum average, Entropy, Difference entropy, Inf…	0.0291135	0.0181621	0.0104869
5	(0, 5, 8, 10, 12)	[0.7, 0.65, 0.633333333333333, 0.633333333333…	0.654167	(Angular second moment, Sum average, Entropy,…	0.0437977	0.0273227	0.0157747
6	(0, 5, 7, 8, 10, 12)	[0.683333333333333, 0.666666666666666, 0.6,…	0.641667	(Angular second moment, Sum average, Sum entr…	0.0550772	0.0343592	0.0198373
7	(0, 5, 7, 8, 9, 10, 12)	[0.666666666666666, 0.65, 0.65, 0.65]	0.654167	(Angular second moment, Sum average, Sum entr…	0.0115685	0.00721688	0.00416667

Fig. 4. Output metric for SFS showing score, std deviation and error of the algorithm

The greedy search algorithms are best fitted to reduce the large number of features that confuses the classifier and degrades its performance. Here, HSEFS algorithm was found the most suitable method for lung cancer dataset in boosting the performance of the random forest classifier shown in Table 1. The table shows the Comparison results of the proposed method with the other implemented methods. Random forest classifier has been applied in all the approaches, which results in 0.99 score for proposed method. The number of best selected features chosen are seven for other feature selection methods, three to seven ranges for the proposed method. Table 2 shows the comparative study of the proposed method and others, obtained the accuracy of 99 % with the proposed algorithm.

Feature subset selection in SFS, cross validation scores, features names, standard deviation and the standard error with the corresponding feature subset can be seen in the output metric shown in Fig. 4. The algorithm adds one feature in feature id and keeps checking the classifier score. The best feature score achieved using this method is 0.654 with 7 feature space. Output metric for SBS shows the working of the algorithm, it evaluates the full feature set and keeps removing the features one by one while measuring the score shown in Fig. 5. The cross validation score obtained with this method is similar to SFS is 0.658. Whereas, After implementing all the feature selection methods, the proposed method (HSEFS) was found the best feature selection method for Lung cancer images. The performance result 0.99 has been achieved which is the best fea-

	feature_idx	cv_scores	avg_score	feature_names	ci_bound	std_dev	std_err
13	(0, 1, 2, 3, 4, 5, 6, 7, 8, 9, 10, 11, 12)	[0.65, 0.566666666666667, 0.666666666666666,...	0.633333	(Angular second moment, contrast, correlation...	0.0626554	0.0390868	0.0225668
12	(0, 1, 2, 3, 5, 6, 7, 8, 9, 10, 11, 12)	[0.65, 0.616666666666667, 0.666666666666666,...	0.65	(Angular second moment, contrast, correlation...	0.0327207	0.0204124	0.0117851
11	(0, 1, 2, 3, 5, 6, 7, 8, 10, 11, 12)	[0.666666666666666, 0.666666666666666, 0.633...	0.645833	(Angular second moment, contrast, correlation...	0.0347056	0.0216506	0.0125
10	(0, 1, 2, 3, 5, 6, 7, 8, 10, 11)	[0.666666666666666, 0.616666666666667, 0.666...	0.65	(Angular second moment, contrast, correlation...	0.0327207	0.0204124	0.0117851
9	(0, 2, 3, 5, 6, 7, 8, 10, 11)	[0.65, 0.633333333333333, 0.666666666666666,...	0.6625	(Angular second moment, correlation, Sum of s...	0.039514	0.0246503	0.0142319
8	(2, 3, 5, 6, 7, 8, 10, 11)	[0.666666666666666, 0.666666666666666, 0.683...	0.666667	(correlation, Sum of squares(variance), Sum av...	0.0188913	0.0117851	0.00680414
7	(2, 3, 5, 6, 7, 8, 11)	[0.666666666666666, 0.65, 0.666666666666666...	0.658333	(correlation, Sum of squares(variance), Sum av...	0.0133582	0.00833333	0.00481125
6	(2, 3, 5, 7, 8, 11)	[0.666666666666666, 0.65, 0.633333333333333,...	0.658333	(correlation, Sum of squares(variance), Sum av...	0.0298698	0.0186339	0.0107583
5	(2, 3, 5, 8, 11)	[0.666666666666666, 0.633333333333333, 0.65,...	0.658333	(correlation, Sum of squares(variance), Sum av...	0.0298698	0.0186339	0.0107583
4	(2, 3, 5, 8)	[0.633333333333333, 0.616666666666667, 0.65,...	0.6375	(correlation, Sum of squares(variance), Sum av...	0.022152	0.0138193	0.00797856
3	(2, 5, 8)	[0.65, 0.633333333333333, 0.65, 0.65]	0.645833	(correlation, Sum average, Entropy)	0.0115685	0.00721688	0.00416667
2	(5, 8)	[0.666666666666666, 0.7, 0.566666666666667,...	0.620833	(Sum average, Entropy)	0.0913352	0.0569783	0.0328964
1	(5,)	[0.516666666666666, 0.6, 0.583333333333334,...	0.583333	(Sum average,)	0.0681136	0.0424918	0.0245327

Fig. 5. Output metric for SBS showing score, std deviation and error of the algorithm

feature_idx	cv_scores	avg_score	feature_names
(0, 1, 2)	[0.9958333333333333]	0.995833	(Angular second moment, Sum average, Sum entr...
(0, 1, 2, 4, 6)	[0.9958333333333333]	0.995833	(Angular second moment, Sum average, Sum entr...
(0, 1, 2, 3, 6)	[0.9958333333333333]	0.995833	(Angular second moment, Sum average, Sum entr...
(0, 1, 2, 3, 5)	[0.9958333333333333]	0.995833	(Angular second moment, Sum average, Sum entr...
(0, 1, 2, 3, 4)	[0.9958333333333333]	0.995833	(Angular second moment, Sum average, Sum entr...
...	...	...	...
(2, 5, 6)	[0.9958333333333333]	0.995833	(Sum entropy, Difference entropy, Info.measure...
(2, 4, 6)	[0.9958333333333333]	0.995833	(Sum entropy, Difference variance, Info.measur...
(2, 4, 5)	[0.9958333333333333]	0.995833	(Sum entropy, Difference variance, Difference ...
(0, 1, 2, 3, 4, 5, 6)	[0.9958333333333333]	0.995833	(Angular second moment, Sum average, Sum entr...
(0, 1, 6)	[0.9916666666666667]	0.991667	(Angular second moment, Sum average, Info.mea...

Fig. 6. Output metric for HSEFS algorithm

ture score in comparison of the other work. The performance metric for HSEFS can be seen in Fig. 6, which describes cross validation score, average score and best selected feature names corresponding to the feature subset combinations. The proposed algorithm selects angular second moment, sum average and sum entropy features as best features in the minimum bound of three.

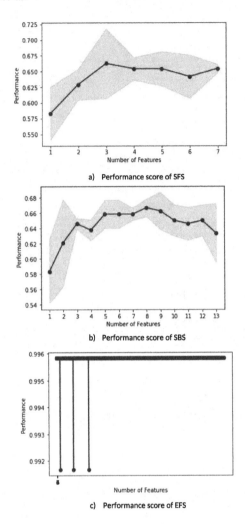

a) Performance score of SFS

b) Performance score of SBS

c) Performance score of EFS

Fig. 7. Performance measures of the proposed feature selection algorithms shown in figure a), b) and c).

5   Conclusion

In this paper, sequential forward selection (SFS), sequential backward selection (SBS) and exhaustive feature selection (EFS) methods along with the proposed algorithm (HSEFS) have been implemented. All the algorithms have been applied on 300 lung CT scan images taken from the LIDC dataset. Texture features were evaluated for all the 300 lung scan nodules, extracted from pylidc library. The SFS method is suitable for large datasets, it consumes less time as compared to other methods. But, the proposed algorithm found suitable for biomedical image classification. The best feature score 0.99 has been achieved from the proposed method, which is a very good for lung cancer cases. Thus, the conclusion is that, the HSEFS feature selection method may be the best solution of feature selection for the biomedical datasets. The computational time is found as the limitation of the work, which may cause complexity in the model.

This paper can be extended further with the evaluation of other feature selection methods for feature selection.

References

1. Bray, F., Ferlay, J., Soerjomataram, I., Siegel, R.L., Torre, L.A., Jemal, A.: Global cancer statistics 2018: GLOBOCAN estimates of incidence and mortality worldwide for 36 cancers in 185 countries. CA: Cancer J. Clin. **68**(6), 394–424 (2018)
2. World Health Organization et al.: What Quantitative and Qualitative Methods Have Been Developed to Measure Community Empowerment at a National Level?, vol. 59. World Health Organization (2018)
3. Thun, M.J., et al.: Lung cancer occurrence in never-smokers: an analysis of 13 cohorts and 22 cancer registry studies. PLoS Med. **5**(9), e185 (2008)
4. Koike, W., Iwano, S., Matsuo, K., Kitano, M., Kawakami, K., Naganawa, S.: Doubling time calculations for lung cancer by three-dimensional computer-aided volumetry: effects of inter-observer differences and nodule characteristics. J. Med. Imaging Radiat. Oncol. **58**(1), 82–88 (2014)
5. Cai, J., Luo, J., Wang, S., Yang, S.: Feature selection in machine learning: a new perspective. Neurocomputing **300**, 70–79 (2018)
6. Iranifam, M.: Analytical applications of chemiluminescence methods for cancer detection and therapy. TrAC Trends Anal. Chem. **59**, 156–183 (2014)
7. Chen, H.-L.: An efficient diagnosis system for detection of Parkinson's disease using fuzzy k-nearest neighbor approach. Expert Syst. Appl. **40**(1), 263–271 (2013)
8. Alharbi, A.: An automated computer system based on genetic algorithm and fuzzy systems for lung cancer diagnosis. Int. J. Nonlinear Sci. Numer. Simul. **19**(6), 583–594 (2018)
9. Joshi, A., Ashish, M.: Analysis of k-nearest neighbor technique for breast cancer disease classification. Int. J. Recent Sci. Res. **8**(8), 1005–19008 (2017)
10. Lakshmanaprabu, S.K., Mohanty, S.N., Shankar, K., Arunkumar, N., Ramirez, G.: Optimal deep learning model for classification of lung cancer on CT images. Future Gener. Comput. Syst. **92**, 374–382 (2019)
11. Alirezaei, M., Niaki, S.T.A., Niaki, S.A.A.: A bi-objective hybrid optimization algorithm to reduce noise and data dimension in diabetes diagnosis using support vector machines. Expert Syst. Appl. **127**, 47–57 (2019)
12. Dhara, A.K., Mukhopadhyay, S., Khandelwal, N.: 3D texture analysis of solitary pulmonary nodules using co-occurrence matrix from volumetric lung CT images. In: Medical Imaging 2013: Computer-Aided Diagnosis, vol. 8670, pp. 867039. International Society for Optics and Photonics (2013)
13. Moradi, P., Gholampour, M.: A hybrid particle swarm optimization for feature subset selection by integrating a novel local search strategy. Appl. Soft Comput. **43**, 117–130 (2016)
14. Liu, X., Ma, L., Song, L., Zhao, Y., Zhao, X., Zhou, C.: Recognizing common CT imaging signs of lung diseases through a new feature selection method based on fisher criterion and genetic optimization. IEEE J. Biomed. Health Inform. **19**(2), 635–647 (2015)

15. Armato III, S.G., et al.: The lung image database consortium (LIDC) and image database resource initiative (IDRI): a completed reference database of lung nodules on CT scans. Med. Phys. **38**(2), 915–931 (2011)
16. Fedorov, A., et al.: Standardized representation of the LIDC annotations using DICOM. Technical report, PeerJ Preprints (2019)
17. Hancock, M.C., Magnan, J.F.: Lung nodule malignancy classification using only radiologist-quantified image features as inputs to statistical learning algorithms: probing the lung image database consortium dataset with two statistical learning methods. J. Med. Imaging **3**(4), 044504 (2016)

Application of Ensemble Techniques Based Sentiment Analysis to Assess the Adoption Rate of E-Learning During Covid-19 Among the Spectrum of Learners

S. Sirajudeen[1] , Balaganesh[2] , Haleema[3] , and V. Ajantha Devi[4(✉)]

[1] Lincoln University College, Petaling Jaya, Malaysia
[2] Faculty of Computer Science and Multimedia, Lincoln University College, Kota Bharu, Malaysia
[3] University of Stirling, RAK Campus, Ras al Khaimah, UAE
[4] AP3 Solutions, Chennai, TN, India

Abstract. The Corona Virus disease (COVID-19) epidemic outbreak leads to worldwide lockdown. Lockdown is enabled to be secure and to keep a correct social distance. According to the Government of India, every university, college and school has been closed defending against this threatening virus. This lockdown time presents an eye-opener for the digital services, for example, use of applications, generating virtual classrooms, online mock testing, online video quizzes, live lectures, deliberations, and document sharing and so on which provides more efficient than ever before. It mostly reveals essential e- learning in education, particularly during this isolation. This paper would assist in identifying attitudes of students' towards e-learning during COVID-19 epidemics using Ensemble Learning-based Sentiment Analysis (ELSA) Algorithm. The study conducts for students studying in different schools, universities and colleges to obtain other data on e-learning involvement during these epidemics. Evaluation metrics, for example, precision, recall, F-score and accuracy are computed and examine for classification performance.

Keywords: Lockdown · E-Learning · Student's Attitude and Sentiment Classification

1 Introduction

On Dec 31, 2019, a pandemic of undiagnosed cause diagnosed in Wuhan, China [1]. On Jan 20, 2020, the World Health Organization (WHO) confirmed the disease transmitted from person to person [2]. The virus that causes COVID-19 is mostly spread by droplets released by a diseased person through coughing, sneezing, or talking [3]. These droplets rapidly spray on surfaces. Humans could be affected by inhaling the virus when they are within a meter of an affected people or by touching a droplets fallen surface, mouth nose, eyes, or using unwashed hands.

The original version of this chapter was revised: The affiliation information of the first author has been corrected as "Lincoln University College, Petaling Jaya, Malaysia". The correction to this chapter is available at https://doi.org/10.1007/978-3-030-82322-1_22

© Springer Nature Switzerland AG 2021, corrected publication 2022
A. Solanki et al. (Eds.): AIS2C2 2021, CCIS 1434, pp. 187–202, 2021.
https://doi.org/10.1007/978-3-030-82322-1_14

The virus, which originates in China, is gradually spreading in extensive to its neighbour nations. A total of 64,270,911 confirmed coronavirus cases have been announced in almost 218 nations and territories worldwide (last updated Dec 2, 2020, by Worldometer [4]). All developed countries suffered major setbacks because of this COVID 19 [5]. The Indian government has announced a lockdown to save its citizens from decreasing the virus's spread [6]. Universities, colleges and Schools have been closures since Mar 17, 2020, to reduce the spread of the COVID-19 epidemic [7–9]. Lockdown is increasing the use of E-learning [10]. Education has entered other stages of attracting students through online classes [11]. Even if universities, colleges and schools are closure, they could pursue their education service through e-learning [12, 13]. Educational institutions offer many online courses, and a few educational institutions provide free classes to engage students [14, 15]. The internet is an essential component of E-learning [16, 17]. At present, a lot of network industries like BSNL, Jio, Vodafone, Airtel, are offering offers to utilize their data. Accompanied by the assist of the internet, you could access online classes on your PC or laptop or even on your smartphone. It is an educational form in which resources, teachers and students interact on the World Wide Web (WWW).

This paper would assist in identifying attitudes of students' towards e-learning during COVID-19 epidemics using Ensemble Learning-based Sentiment Analysis (ELSA) Algorithm. Model is trained based on conventional machine learning methods for classification like Support Vector Machine (SVM), K-Nearest Neighbor (KNN), Naive Bayes (NB), Artificial Neural Network (ANN), and Decision Tree (DT). The system is executed based on an ensemble of these machine learning methods for sentiment analysis. The study conducts for students studying in different schools, universities and colleges to obtain other data on e-learning involvement during these epidemics. It could be a learning model in educational institutions to improve students' knowledge and skills via digital technologies. The government and the education sector should present superior infrastructure for E-learning for the development of students. The paper's remaining section is structured as follows—the related work about sentiment analysis reviewed in Section 2. Ensemble learning for sentiment analysis, as explained in Section 3. The results of the experiments are discussed in Section 4, followed by Section 5, which concludes the paper.

2 Related Work

Kastrati et al. [18] present a structure for auto-examining student feedback expressed in the analysis. In particular, the structure depends on feature-level perception reviews and goals to spontaneously discover the opinion or sentiment polarity towards a provided feature associated with MOOC [19]. The presented structure uses a weakly-supervised annotation of MOOC-associated features and transmits a weakly supervised signal to efficiently discover the feature types debated in anonymous student analysis. As a result, it notably decreases the necessity for manually annotated information, which is a significant bottleneck to entire in-depth learning technologies. A huge-scale actuality educational database, including an analysis of approximately 105k of students gathered from Coursera [20]. A database of 5989 student comments in conventional classroom settings utilized to execute the experiments. The authors concluded that structuring guides to more precise outcomes than costly and labour intensive sentiment reviews technologies that rely massively on manually labelled information.

Fang et al. [21] suggested several strategic sentiment analysis techniques accompanied by semantic ambiguity. The authors concluded that this mixed emotion analysis technique could attain better performance. At the Autonomous University of Madrid, Spain, Cobos et al. [22] have planned and created a tool for using NLP methods to examine online courses contents [23]. The instructional materials of these courses and the participation of their students enhance teaching and learning practices. This tool is named edX-CAS (edX MOOCs - Content Analyzer System). Cobos et al. [22] present a comprehensive explanation of the tool, its functions, and its NLP procedures, which assist emotional research for personal and polarization discovery. Furthermore, the authors provided a survey of recent study in the domain of appliances of the NLP in improving the learning and teaching experiences in MOOCs.

Al-Moslmi et al. [24] provided the outcomes of an extensive structured literature review of the technologies used in a cross-field opinion analysis. The authors concentrated on learning's issued while the period of 2010–2016. From their research of those tasks, it is transparent that there is no pure problem-solving method. Therefore, one of the objectives of this survey is to develop a resource in the shape of a summary of technologies that try to resolve cross-field opinion analysis to help analyzers create novel and exact technologies in the upcoming. An advanced word portrayal system was presented by Xu et al. [25], which combines the participation of sentiment data to the conventional TF-IDF system and creates weighted word vectors [26]. Weighted word vectors are entered into bidirectional long short term memory (BiLSTM) to catch contextual data efficiently, and comment vectors best portrayed [25]. The emotional pathway of the concept derived from the Feed-Forward neural network classifier [27]. Under similar states, the presented sentiment analysis technique compared accompanied by the emotional research methods of NB, LSTM, CNN and RNN. The authors concluded that the presented emotional research technique had higher F1 score, recall and accuracy.

Xu et al. [28] presented an Emotion analysis technique based on essential data topics for big data. This technique combines topic semantic data with text representation via a neural network method. The focusing algorithm launched in the neural network and the environmental awareness vector initiated to compute each term's weight—additionally, the Sentiment Dictionary tag system utilized to retrieve training data to modify the method further [29]. The authors concluded that the presented method could enhance the precise of the sentiment analysis outcomes.

Zhang et al. [30] created a conversation database for authors and made it publicly known as ScenarioSA for collaborative emotion analysis. Authors manually name 2,214 multi-turn English discussions gathered from different websites that present online transmission services. Compared to previous sentimental databases, ScenarioSA (1) is no extensive restricted to a particular field; however, it covers a broad range of views and topics; (2) reports the relationship among the two speakers of each discussion; (3) mirrors the emotional developments of each speaker during the discussion. Eventually, the authors suggest an expansion of collaborative attention systems that can method the communications, contrast different powerful opinion analysis algorithms on ScenarioSA, and demonstrate the necessity of new collaborative opinion analysis methods and the possibility of ScenarioSA to assist the evolution of such methods.

Aziz et al. [31] offer a technique called Contextual Analysis (CA), which builds a connection between sources and words structured in a tree form known as the Hierarchical Knowledge Tree (HKT). Later, the Tree Differences Index (TDI) and the Tree Similarity Index (TSI), a formula derived from the tree form, were suggested to discover the similarities and differences between the real and the train data set. The regression analysis of the datasets unveils that there is a remarkable positive connection between SML and TSI precisions. Chen et al. [32] initiated the weakly supervised multimodal deep learning (WS-MDL) system toward substantial and measurable emotional prediction. WS-MDL restructures and reconstructs neural networks from inexpensively existing and noiseless "weak" smiley labels. Mostly, a probabilistic graphical model has been initiated to filtrate label noise, catch natural dependencies, study to simulate multimodal descriptors simultaneously, and predict the reliability of label noise. Large-scale ratings managed on a Long and short scales, actual-world micro-blog emotional database, crawled from Sina Weibo [33]. The authors have verified the advantages of the WS-MDL system by demonstrating the degree of its optimal effectiveness in many sophisticated and other possible methods.

Hasan et al. [34] presented to adopt a combined method that includes an opinion analysis that contains machine learning. Furthermore, the author presents comparisons of the methods of opinion analysis in the analysis of governmental perspectives, and supervised machine learning methods, such as support vector machine (SVM) [35] and Naïve Bayes. Imran et al. [36] examined the response of citizens from various traditions to the new coronavirus and the public perception of the following activities taken by multiple nations. Deep long short-term memory (LSTM) methods [37] utilized to estimate the sentimental polarization and emotion from takeout tweets trained to attain sophisticated precision in the Sentiment140 dataset. A utilize of emoticons demonstrated a distinctive and new method to check the supervised deep learning patterns in tweets take out from Twitter. Jin et al. [38] presented a long short term memory (LSTM), and a multi-scale convolutional neural network (CNN) based multi-task learning method (MTL-MSCNN-LSTM) for multi-scale multi-task sentiment classification. This method uses and precisely manipulates the local and global features of texts of various sizes to represent sentences and patterns. The multi-task learning structure enhances encryption standard, while at the same time strengthening the outcomes of emotion classification—six various kinds of product review datasets used in the evaluation. The proposed method can be used accurately and as an F1 score to assess the performance of the proposed MTL-MSCNN-LSTM method compared to models, for example, single-task learning and LSTM encoding. The proposed method works better than most previous methods.

Feizollah et al. [39] explained how Twitter information is excerpted, and the sentiment of tweets on a specific subject computed. The authors focused on the tweets of two Halal Travel, halal products, and Halal Cosmetics. Twitter information (over ten years) excerpted based on the Twitter explore method, and an algorithm utilized to filtrate the information. After that, an experiment managed to compute and examine the appearance of tweets based on deep learning methods. Additionally, recurrent neural networks (RNN) and long-term memory (LSTM) enhance precision and develop predictive methods. In [40] the authors discussed various research contemplate on educational data mining and several useful algorithm were analyzed through this study. The authors studied the association of self-esteem and students' performance in academics [41]. Firoz khan et al. analyzed a good area of using machine learning through a digital DNA sequencing engine for ransomware detection and detecting malicious URLs using binary classification through ada boost algorithm. These concepts were also analyzed for the possible inclusion in the education domain as well [42, 43].

3   Methodology

This methodology would assist in identifying attitudes of students' towards e-learning during COVID-19 epidemics using Ensemble Learning-based Sentiment Analysis (ELSA) Algorithm. It is trained based on conventional machine learning methods for classification like Support Vector Machine (SVM), K-Nearest Neighbor (KNN), Naive

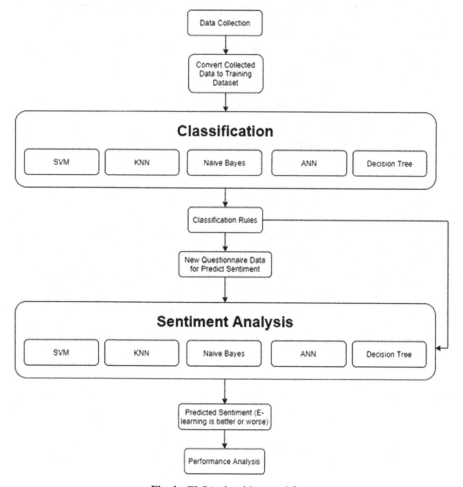

Fig. 1. ELSA algorithm workflow

Bayes (NB), Artificial Neural Network (ANN), and Decision Tree (DT). This section describes the entire research work method from data collection to performance analysis, as demonstrated in Fig. 1.

Algorithm 1 explained the proposed ELSA algorithm. A questionnaire with ten different questions about E-learning is gathered from different students to collect a meaningful data set—these questionnaire data collected from Google Forms. After collecting the data from Google Forms, data converted as a training dataset performed (Step 1). Furthermore, the dataset undergoes training with ensemble ML models as SVM, KNN, ANN, NB and DT. It provides some classification rules (Step 2–6).

Furthermore, new questionnaire data predicts sentiment based on these rules (Step 7–12). Next, the ESLA algorithm concludes the most frequent term as predicted sentiment (Step 13–15). Finally, the ESLA algorithm compares this ensemble learning classification algorithms based on precision, recall, f-measure and accuracy. Models are further tuned to improve the performances using hyperparameters, and finally, the results of all ML models compared.

Algorithm 1: Ensemble Learning-based Sentiment Analysis (ELSA)	
Input	The questionnaire with responses collections (QR), a questionnaire with the response to predict (TeD)
Output	Predicted Sentiment (PS) about E-learning (Better or Worse)
Step 1	TrD = Convert QR to Training Dataset
Step 2	SVMC = Classification(TrD, "SVM") // Algorithm 2
Step 3	KNNC = Classification(TrD, "KNN")
Step 4	NBC = Classification(TrD, "NB")
Step 5	ANNC = Classification(TrD, "ANN")
Step 6	DTC = Classification(TrD, "DT")
Step 7	For Questionnaire Q from TeD
Step 8	SVMPS = SentimentAnalysis(Q, SVMC) // Algorithm 3
Step 9	KNNPS = SentimentAnalysis(Q, KNNC)
Step 10	NBPS = SentimentAnalysis(Q, NBC)
Step 11	ANNPS = SentimentAnalysis(Q, ANNC)
Step 12	DTPS = SentimentAnalysis(Q, DTC)
Step 13	ArrayList FR = new ArrayList();
Step 14	FR.add(SVMPS); FR.add(KNNPS); FR.add(NBPS); FR.add(ANNPS); FR.add(DTPS);
Step 15	PS = Get Most Frequent Term from FR // Predicted Sentiment
Step 16	End For
Step 17	Compare Ensemble Learning Classification algorithms based on Performance Metrics

3.1 Data Collection

This analysis is using primary data. Data recently collected from students pursuing UG degrees at various universities and colleges. The online questionnaire is prepared to gather data through Google Forms. The analysis period is November 2020. The study conducted for students learning at different schools, universities and colleges to profit additional data on the involvement of e-learning during these epidemics. Data collection through a questionnaire prepared by the authors is an economical way to obtain information. It is easily accessible and offers comprehensive protection with minimal effort.

3.2 Classification

This section classifies the training dataset with ensemble ML models as SVM, KNN, ANN, NB and DT. Algorithm 2 explain the ML classification algorithm using WEKA. First, this algorithm creates a new FileReader given the name of the Training dataset file (TrD) to read from (Step 1). It reads an ARFF format TrD from a reader and assigns a weight of one to each instance (Step 2). Followed by, it sets the class index of the TrD. In TrD, the last attribute act as a class attribute (Step 3). Next, it generates a machine learning algorithm classifier for TrD using WEKA (Step 4–5). It provides some classification rules for sentiment analysis. Followed by, this algorithm evaluates the classifier on a training set of instances (Step 6–7).

Algorithm 2: Classification		
Input	:	Training Dataset (TrD), ML Algorithm (MLA)
Output	:	Classification Rules (CR)
Step 1	:	FileReader trainreader = new FileReader(TrD);
Step 2	:	Instances train = new Instances(trainreader);
Step 3	:	train.setClassIndex(train.numAttributes()-1);
Step 4	:	MLA mla = new MLA();
Step 5	:	CR = mla.buildClassifier(train);
Step 6	:	Evaluation eval1 = new Evaluation(train);
Step 7	:	eval1.evaluateModel(mla, train);

3.3 Sentiment Analysis

This section predicts sentiment for a new questionnaire with a response based on Classification rules generated after classification. Algorithm 3 explains the Sentiment Analysis algorithm. This algorithm first converts the Questionnaire to the ARFF format testing dataset (Step 1). This algorithm creates a new FileReader given the name of the Testing dataset file (TeD) to read from (Step 2). Then it reads an ARFF format TeD from a reader and assigns a weight of one to each instance (Step 3). Followed by, it sets the class index of the TeD. In TeD, the last attribute act as a class attribute (Step 4). Then, each instance classifies based on CR and predict class value (Step 5–10). Followed by, this algorithm evaluates the classifier on a testing set of instances (Step 11–12).

Algorithm 3: Sentiment Analysis		
Input	:	Questionnaire (Q), Classification Rules (CR)
Output	:	Predicted Sentiment (PS)
Step 1	:	TeD = Convert Q to Testing Dataset
Step 2	:	FileReader testreader = new FileReader(TeD);
Step 3	:	Instances test = new Instances(testreader);
Step 4	:	test.setClassIndex(test.numAttributes()-1);
Step 5	:	for (int i = 0; i < test.numInstances(); i++)
Step 6	:	{
Step 7	:	double clsLabel = CR.classifyInstance(test.instance(i));
Step 8	:	test.instance(i).setClassValue(Math.abs((int)clsLabel));
Step 9	:	PS = test.instance(i).getClassValue();
Step 10	:	}
Step 11	:	Evaluation eval1 = new Evaluation(test);
Step 12	:	eval1.evaluateModel(CR, test);

4   Experimental Results and Discussions

4.1   Questionnaire Method for Data Collection

To understand students' attitudes on the execution of e-learning during Covid-19 epidemics, different students answer the following questions showed in Table 1:

Table 1. Questionnaire

Q. No	Questions
1	Do you like e-learning?
2	Does e-learning boost your skills?
3	Does e-learning save your time and money?
4	Is e-Learning is a way to provide quick delivery of lessons?
5	Is e-Learning could solve teacher scarcity?
6	Is e-learning useful for your isolated time?
7	Does e-learning expand your knowledge?
8	Is E-learning has less impact on the environment?

(continued)

Table 1. (*continued*)

Q. No	Questions
9	Is E-learning helps to retain information for a long time?
10	Does online learning is better than traditional learning?

This paper collected questionnaire data from Google Forms. After collecting the data from Google Forms, data converted as a training dataset performed. While training, the ELSA algorithm applied ensemble ML models as SVM, KNN, ANN, NB and DT. It provides some classification rules. Furthermore, new questionnaire data give to predict sentiment based on these rules. Next, the ESLA algorithm concludes the most frequent term as the predicted sentiment. Figure 2 shows sentiment analysis results about E-learning among 850 student Questionnaire responses.

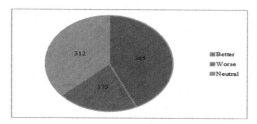

Fig. 2. Sentiment analysis results about E-learning among 850 student Questionnaire responses

Figure 2 shows that E-learning has become very famous among students worldwide, especially the lockdown time because of the COVID-19 epidemic. Finally, the ESLA algorithm compares this ensemble learning classification algorithms based on precision, recall, f-measure and accuracy.

1) **Precision**

Precision utilized to reduce the number of false positives. It verifies how many times the classifier forecasts positive outcomes. It computed as the amount of true positive forecasts divided by the total amount of positive forecasts, as demonstrated in Eq. 1. It is the capability to return merely relevant instances of a classification algorithm. It is also called a positive predicted value.

$$\text{Precision} = \text{TP}/\text{TP} + \text{FP} \tag{1}$$

Where TP is True Positive, and FP is False Positive. Table 2 shows the precision of executed Ensemble Learning techniques.

Table 2. The precision of executed ensemble learning techniques

Algorithm	Precision
SVM	0.955
KNN	0.990
Naïve Bayes	0.852
ANN	0.859
DT	0.832

Figure 3 shows compared precision of SVM, KNN, Naïve Bayes, ANN and DT algorithms.

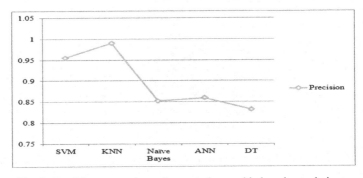

Fig. 3. Precision comparison of executed ensemble learning techniques

Figure 3 interpret KNN algorithm provides the best performance in terms of precision. Followed by, the SVM algorithm offers better performance than others. Naïve Bayes, ANN and DT performance are low in terms of precision.

2) **Recall**

It computed the sensitivity and computed as the amount of true positive forecasts divided by the total amount of positives, as demonstrated in Eq. 2. Worst Sensitivity is 0, and the Best is 1. A classification method can discover entire relevant instances. It is also called the true positive rate.

$$Recall = TP / TP + FN \qquad (2)$$

Where TP is True positive, and FN is False Negative. Table 3 shows the recall of executed Ensemble Learning techniques.

Table 3. Recall of executed ensemble learning techniques

Algorithm	Recall
SVM	0.704
KNN	0.990
Naïve Bayes	0.648
ANN	0.990
DT	0.742

Figure 4 Recall of SVM, KNN, Naïve Bayes, ANN and DT algorithms.

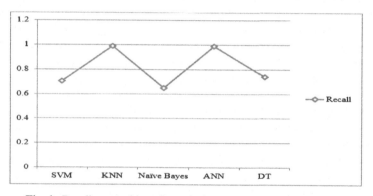

Fig. 4. Recall comparison of executed ensemble learning techniques

Figure 4 interpret KNN and ANN algorithm provides the best performance in terms of recall. Followed by, DT algorithm offers better performance than others. Naïve Bayes and SVM performance are low in terms of recall.

3) **F-Score**

F-Score is a measurement of test accuracy is explained as the weighted harmonic mean of the test's recall and precision. A single metric merges precision and recall based on the harmonic mean as demonstrated in Eq. 3.

$$F - Score = 2 * (Precision * Recall) / (Precision + Recall) \qquad (3)$$

Table 4 shows the F-Score of executed Ensemble Learning techniques.

Table 4. F-Score of executed ensemble learning techniques

Algorithm	F-Score
SVM	0.811
KNN	0.990
Naïve Bayes	0.736
ANN	0.924
DT	0.784

Figure 5 shows compared F-Score of SVM, KNN, Naïve Bayes, ANN and DT algorithms.

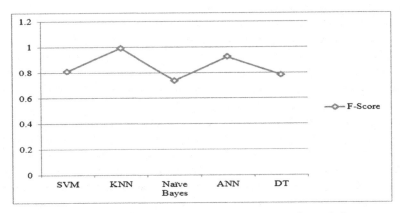

Fig. 5. F-Score comparison of executed ensemble learning techniques

Figure 5 interpret KNN algorithm provides the best performance in terms of F-Score. Followed by, the ANN algorithm offers better performance than others. Naïve Bayes, SVM and DT performance are low in terms of F-Score.

4) Accuracy

Accuracy is a measurement utilized for ML methods to determine the winning method for discovering patterns and relationships among variables in a dataset based on supervised information. Accuracy could compute, as demonstrated in Eq. 4.

$$Accuracy = TP + TN / TP + FP + TN + FN \tag{4}$$

True forecasts of the method are True Negatives and True positives. All other forecasts are False Negatives (FN), False Positives (FP), True Negatives (TN) and True Positives (TP). Table 5 shows the accuracy of executed Ensemble Learning techniques.

Table 5. Accuracy of executed ensemble learning technique

Algorithm	Accuracy
SVM	0.847
KNN	0.990
Naïve Bayes	0.803
ANN	0.937
DT	0.811

Figure 6 Accuracy comparison of SVM, KNN, Naïve Bayes, ANN and DT algorithms.

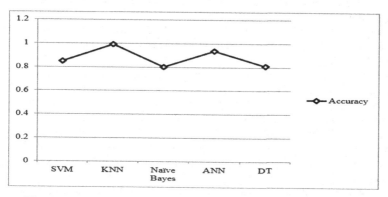

Fig. 6. Accuracy comparison of executed ensemble learning techniques

Figure 6 interpret KNN algorithm provides the best performance in terms of accuracy. Followed by, the ANN algorithm offers better performance than others. Naïve Bayes, SVM and DT performance are low in terms of accuracy.

5 Conclusion

This paper helps to discover attitudes of students' towards e-learning during COVID-19 epidemics using Ensemble Learning-based Sentiment Analysis (ELSA) Algorithm. It is trained based on conventional machine learning [44] methods for classification like Support Vector Machine (SVM), K-Nearest Neighbor (KNN), Naive Bayes (NB), Artificial Neural Network (ANN), and Decision Tree (DT). The system is executed based on an ensemble of these machine learning methods for sentiment analysis. This paper presented experimental results comparison between these ML techniques on different performance metrics to predict sentiment analysis. It concluded that compared with other ML techniques, the KNN algorithm provides the best performance. This paper concluded that "COVID-19 is negative; however, it is positive for online learning".

References

1. Zheng, N., et al.: Predicting covid-19 in china using hybrid AI model. IEEE Trans. Cybern. 8 May 2020
2. Chamola, V., Hassija, V., Gupta, V., Guizani, M.: A comprehensive review of the COVID-19 pandemic and the role of IoT, drones, AI, blockchain, and 5G in managing its impact. IEEE Access **8**, 90225–90265 (2020)
3. Alsaeedy, A.A., Chong, E.K.: Detecting regions at risk for spreading COVID-19 using existing cellular wireless network functionalities. IEEE Open J. Eng. Med. Biol. **15**(1), 187–189 (2020)
4. COVID-19 CORONAVIRUS PANDEMIC. https://www.worldometers.info/coronavirus/
5. Kim, R.Y.: The impact of COVID-19 on consumers: preparing for digital sales. IEEE Eng. Manag. Rev. 23 April 2020
6. Singh, R.K., et al.: Short-term statistical forecasts of COVID-19 infections in India. IEEE Access. **8**(8), 186932–186938 (2020)
7. Nassr, R.M., Aborujilah, A., Aldossary, D.A., Aldossary, A.A.: Understanding education difficulty during COVID-19 lockdown: reports on Malaysian University students' experience. IEEE Access. **12**(8), 186939–186950 (2020)
8. Ocaña, J.M., Morales-Urrutia, E.K., Pérez-Marín, D., Pizarro, C.: Can a learning companion be used to continue teaching programming to children even during the COVID-19 pandemic? IEEE Access. **27**(8), 157840–157861 (2020)
9. Nugroho, R.A., Basari, A., Suryaningtyas, V.W., Cahyono, S.P.: University students' perception of online learning in Covid-19 pandemic: a case study in a translation course. In: 2020 International Seminar on Application for Technology of Information and Communication (iSemantic) 19 September 2020, pp. 225–231. IEEE (2020)
10. Islam, M.N., Islam, I., Munim, K.M., Islam, A.N.: A review on the mobile applications developed for COVID-19: an exploratory analysis. IEEE Access **7**(8), 145601–145610 (2020)
11. Ahmed, N., et al.: A survey of Covid-19 contact tracing apps. IEEE Access **20**(8), 134577–134601 (2020)
12. Horry, M.J., et al.: COVID-19 detection through transfer learning using multimodal imaging data. IEEE Access **14**(8), 149808–149824 (2020)
13. Ramchandani, A., Fan, C., Mostafavi, A.: DeepCOVIDNet: an interpretable deep learning model for predictive surveillance of Covid-19 using heterogeneous features and their interactions. IEEE Access **28**(8), 159915–159930 (2020)
14. Al-Rahmi, W.M., et al.: Use of e-learning by university students in Malaysian higher educational institutions: a case in Universiti Teknologi Malaysia. IEEE Access **7**(6), 14268–14276 (2018)
15. Moubayed, A., Injadat, M., Nassif, A.B., Lutfiyya, H., Shami, A.: E-learning: challenges and research opportunities using machine learning & data analytics. IEEE Access **23**(6), 39117–39138 (2018)
16. Kanwal, F., Rehman, M.: Factors affecting e-learning adoption in developing countries–empirical evidence from Pakistan's higher education sector. IEEE Access **9**(5), 10968–10978 (2017)
17. Al-Tarabily, M.M., Abdel-Kader, R.F., Azeem, G.A., Marie, M.I.: Optimizing dynamic multi-agent performance in an E-learning environment. IEEE Access **15**(6), 35631–35645 (2018)
18. Kastrati, Z., Imran, A.S., Kurti, A.: Weakly supervised framework for aspect-based sentiment analysis on students' reviews of MOOCs. IEEE Access **8**(8), 106799–106810 (2020)
19. Rasheed, R.A., Kamsin, A., Abdullah, N.A., Zakari, A., Haruna, K.: A systematic mapping study of the empirical MOOC literature. IEEE Access **30**(7), 124809–124827 (2019)
20. Wu, B., Zhou, Y.: The impact of MOOC instructor group diversity on review volume and rating—coursera specialization as an example. IEEE Access **8**(8), 111974–111986 (2020)

21. Fang, Y., Tan, H., Zhang, J.: Multi-strategy sentiment analysis of consumer reviews based on semantic fuzziness. IEEE Access **12**(6), 20625–20631 (2018)
22. Cobos, R., Jurado, F., Blázquez-Herranz, A.: A content analysis system that supports sentiment analysis for subjectivity and polarity detection in online courses. IEEE Revista Iberoamericana de Tecnologías Del Aprendizaje **14**(4), 177–187 (2019)
23. Bahja, M., Safdar, G.A.: Unlink the link between COVID-19 and 5G networks: an NLP and SNA based approach. IEEE Access **18**(8), 209127–209137 (2020)
24. Al-Moslmi, T., Omar, N., Abdullah, S., Albared, M.: Approaches to cross-domain sentiment analysis: a systematic literature review. IEEE Access **31**(5), 16173–16192 (2017)
25. Xu, G., Meng, Y., Qiu, X., Yu, Z., Wu, X.: Sentiment analysis of comment texts based on BiLSTM. IEEE Access **9**(7), 51522–51532 (2019)
26. Zhu, Z., Liang, J., Li, D., Yu, H., Liu, G.: Hot topic detection based on a refined TF-IDF algorithm. IEEE Access **31**(7), 26996–27007 (2019)
27. Jin, L., Huang, Z., Li, Y., Sun, Z., Li, H., Zhang, J.: On modified multi-output Chebyshev-polynomial feed-forward neural network for pattern classification of wine regions. IEEE Access **7**(7), 1973–1980 (2018)
28. Xu, G., Yu, Z., Chen, Z., Qiu, X., Yao, H.: Sensitive information topics-based sentiment analysis method for big data. IEEE Access **8**(7), 96177–96190 (2019)
29. Li, Z., Li, R., Jin, G.: Sentiment analysis of danmaku videos based on Naïve Bayes and sentiment dictionary. IEEE Access **8**(8), 75073–75084 (2020)
30. Zhang, Y., Zhao, Z., Wang, P., Li, X., Rong, L., Song, D.: ScenarioSA: a dyadic conversational database for interactive sentiment analysis. IEEE Access **12**(8), 90652–90664 (2020)
31. Aziz, A.A., Starkey, A.: Predicting supervised machine learning performances for sentiment analysis using contextual-based approaches. IEEE Access **10**(8), 17722–17733 (2019)
32. Chen, F., Ji, R., Su, J., Cao, D., Gao, Y.: Predicting microblog sentiments via weakly supervised multimodal deep learning. IEEE Trans. Multimedia **20**(4), 997–1007 (2017)
33. Lei, K., et al.: Understanding user behaviour in sina weibo online social network: a community approach. IEEE Access **20**(6), 13302–13316 (2018)
34. Hasan, A., Moin, S., Karim, A., Shamshirband, S.: Machine learning-based sentiment analysis for Twitter accounts. Math. Comput. Appl. **23**(1), 11 (2018)
35. Al-Smadi, M., Qawasmeh, O., Al-Ayyoub, M., Jararweh, Y., Gupta, B.: Deep recurrent neural network vs support vector machine for aspect-based sentiment analysis of Arabic hotels' reviews. J. Comput. Sci. **1**(27), 386–393 (2018)
36. Imran, A.S., Daudpota, S.M., Kastrati, Z., Batra, R.: Cross-cultural polarity and emotion detection using sentiment analysis and deep learning on COVID-19 related tweets. IEEE Access **28**(8), 181074–181090 (2020)
37. Yu, L., Chen, J., Ding, G., Tu, Y., Yang, J., Sun, J.: Spectrum prediction based on Taguchi method in deep learning with long short-term memory. IEEE Access **7**(6), 45923–45933 (2018)
38. Jin, N., Wu, J., Ma, X., Yan, K., Mo, Y.: Multi-task learning model based on multi-scale CNN and LSTM for sentiment classification. IEEE Access **22**(8), 77060–77072 (2020)
39. Feizollah, A., Ainin, S., Anuar, N.B., Abdullah, N.A., Hazim, M.: Halal products on Twitter: data extraction and sentiment analysis using a stack of deep learning algorithms. IEEE Access **17**(7), 83354–83362 (2019)
40. Jayanthi, M.A., Kumar, R.L., Surendran, A., Prathap, K.: Research contemplate on educational data mining. In: 2016 IEEE International Conference on Advances in Computer Applications (ICACA), Coimbatore, pp. 110–114 (2016). https://doi.org/10.1109/ICACA.2016.7887933
41. Jayanthi, M., Kumar, R., Swathi, S.: Investigation on association of self-esteem and students' performance in academics. Int. J. Grid Util. Comput. **9**, 211 (2018). https://doi.org/10.1504/IJGUC.2018.093976

42. Khan, F., Ncube, C., Ramasamy, L.K., Kadry, S., Nam, Y.: A digital DNA sequencing engine for ransomware detection using machine learning. IEEE Access **8**, 119710–119719 (2020). https://doi.org/10.1109/ACCESS.2020.3003785
43. Khan, F., et al.: Detecting malicious URLs using binary classification through ADA boost algorithm. Int. J. Electrical Comput. Eng. **10**, 2088–8708 (2020)
44. Ajantha Devi, V.: Supervised learning approach to object identification and recognition. Int. J. Pure Appl. Math. **119**(10), 463–470 (2018)

A Hybrid Mathematical Model Using DWT and SVM for Epileptic Seizure Classification

Jigyasa Nayak$^{(\boxtimes)}$ ⓘ, Jasdeep Kaur, and Akash Tayal ⓘ

Department of Electronics and Communication Engineering, Indira Gandhi Delhi Technical University for Women, New Delhi, India
jasdeep@igdtuw.ac.in

Abstract. This paper illustrates an automatic seizure detection framework that is based on discrete wavelet transforms (DWT), non-linear and statistical features, and support vector machines (SVMs). Electroencephalogram (EEG) signals possess non-linear and rhythmic properties in different frequency bands. Thus, the non-linear features are widely used to advance epileptic seizure detection models and achieve promising results. This research work aims to consider multiple non-linear features so that if the information is missed by one non-linear measure, it can be captured by another. The non-linear features are further combined with the statistical features as statistical features help get better epileptic seizure classification accuracy. All features are calculated on D(2), D(3), D(4), D(5), and A(5) wavelet sub-bands, then combined into a single vector and classified using SVMs. The intended approach's accomplishment is assessed with respect to terms sensitivity, specificity and accuracy, tested at the University of Bonn and Neurology and Sleep Centre datasets.

Keywords: Epileptic seizure · Epilepsy · Discrete wavelet transforms · Non-linear methods · Support vector machines

1 Introduction

Epilepsy is an ailment that is illustrated by recurring seizures. As per the WHO report [1], it affects more than 50 million people worldwide and 2.4 million people are detected with epilepsy yearly. All ages and genders are affected by the problem of epileptic seizures. The occurrence rate for the same is very high. The quality-of-life impairments related to epilepsy as compared to other chronic ailments e.g. hypertension, diabetes and heart disease [2] is very high. Moreover, epileptic seizures are quite unpredictable [3]. The report [1] conveys that 80% of the people who are having epilepsy lives in developing countries and more than 70% do not have the facilities of treatments. Such a high ratio is insufficient knowledge of Anti-Epileptic Drugs (AEDs), poverty and shortage of trained professionals [4–7]. The data suggest that at primary levels ratio of neurologist vs population 1:1,000,000 in India compared to 1: 26000 for USA, India faces a huge shortage of neurologists [8]. The Epilepsy treatment index of people vs place varies from 22% (urban) to 90% (villages) [4]. The biggest challenge is reducing the misdiagnosis;

© Springer Nature Switzerland AG 2021
A. Solanki et al. (Eds.): AIS2C2 2021, CCIS 1434, pp. 203–218, 2021.
https://doi.org/10.1007/978-3-030-82322-1_15

the reported misdiagnosis rate varies between 2% and 71% [9]. Patients suffering from epileptic seizures and their families are entitled to diagnosis, prognosis, and precise and specific management [10].

Video- (EEGs) still serves as the most efficient method for detecting epileptic seizure. [11]. EEG gives appropriate information about the initiation of the seizure because of its fine temporal resolution. EEG provides both electrophysiological and behavioral information. An epileptiologist inspects the EEG recordings to classify the epileptic seizure. To manually inspect the video-EEG of patients makes the job very tedious and places a burden on clinical staff. Therefore, to cope with the above difficulties, an automated solution is required to accurately detect epilepsy for patients to start AEDs therapy and reduce the occurrence and complications related to seizures. EEG signal records the seizure and is widely adopted for the epilepsy diagnosis [12]. EEG is a non-stationary and non-linear signal that possesses rhythmic property in five frequency bands, named as delta (0.1–4 Hz), theta (4–8 Hz), alpha (8–12 Hz), beta (12–30 Hz) and gamma (30–45 Hz) [13]. Due to non-stationary and non-linear characteristics, DWT (DWT) and non-linear feature extraction techniques are widely adopted for epilepsy detection [14–19]. The benefit of using DWT is that it offers optimal time-frequency resolution across all frequency ranges [20], and is very helpful in analyzing and helps analyze non-stationary signals. Moreover, non-linear features extracted from the wavelet coefficients leads to good classification accuracy [14]. The advantage of using non-linear features is that it has a very high discrimination ability to segregate normal and epileptic EEG signals [16, 21–24].

In the past, machine learning models considering non-linear features are being extensively used for seizure detection and have attained the highest level of accuracy. In this paper, support vector machines (SVMs) based automated epileptic seizure detection model is intended. The machine learning models consider non-linear and statistical features to make the classification. The authors have considered multiple non-linear features, such as fractal dimension, approximate entropy (ApEn) and Hjorth parameters so that the information missed by one non-linear measure can be captured by other non-linear measures, which may result in better accuracy. The non-linear features are further combined with the statistical features as statistical features helped in achieving better seizure classification accuracy [18]. The accomplishment of the intended approach w.r.t sensitivity, specificity, and accuracy is tested at the University of Bonn, and Neurology and Sleep Centre dataset.

2 Literature Review

Many algorithms based on non-linear features have been developed in recent years to develop efficient epileptic seizure classification models. Most of these algorithms addressed the problem of epileptic seizure classification using machine learning algorithms. Some authors [14, 15, 25–27] have considered a single non-linear feature, while some authors [18, 19, 21, 28–30] have considered multiple non-linear features to design the epileptic classification model. Entropy is regarded as a well-known non-linear feature for the development of epileptic seizure classifier and has been extensively used [14, 18, 19, 21, 25, 26, 29, 30]. The author in [14] used ApEn as a feature to segregate normal and epileptic signals. The author can classify the epileptic EEG signals with 96% accuracy when features are calculated from DWT coefficients. The accuracy was just 73% when the features are calculated from the time series EEG signals. Authors in [21] have also applied entropy estimators to categorize normal and epileptic signals and distinguished epileptic EEG data with greater than 95% certainty. Authors in [27] have considered fractal dimension features and machine learning model to classify the seizure data. The authors have achieved 100% sensitivity for most of the cases, while the accuracy is even 94.5% with some cases. Therefore, to consider one non-linear parameter might be misleading [31]. Different non-linear measures give different kinds of information on the same signal [32]. For instance, the fractal dimension provides the measurement of the EEG signals' complexity and chaotic nature and has good discrimination capability [33]. The ApEn also provides a good measurement of the irregularity of EEG signals. The ApEn value drops abruptly during epileptic seizures due to neurons' synchronous discharge [22]. For this reason; many authors [18, 29, 30] have considered multiple non-linear measures to minimize the influence of one non-linear measure on another. However, the previous studies lack conclusive remarks on the qualitative nature of the non-linear features. Though the usage of non-linear features helped achieve promising results for seizure classification, still it is an open research area on how to make use of full potential of non-linear measures to improvise epileptic seizure classification accuracy.

3 Background

3.1 Discrete Wavelet Transforms

Wavelet transform is capable of giving time-frequency representation as time localization of spectral components is of great importance. Wavelets can be broadly categorized, such as Continuous Wavelet Transform (CWT) and DWT. These transforms differ in how the wavelets are scaled and shifted. The CWT analysis is comparatively an expensive task because it involves the computation of wavelet coefficients at each feasible scale value. DWT is more efficient as the scales, and respective shifts are calculated on the base powers of two, also known as dyadic scales. Mathematically CWT and DWT are represented as:

$$\text{CWT}(a, b) = \int_{-\infty}^{\infty} x(t) \frac{1}{\sqrt{|a|}} \psi\left(\frac{t-b}{a}\right) \tag{1}$$

The scaling, shifting parameters, and wavelet function are represented by a, b, and ψ, correspondingly.

$$\text{DWT}(m, n) = \frac{1}{\sqrt{|2^m|}} \int_{-\infty}^{\infty} x(t) \varphi\left(\frac{t - 2^m n}{2^m}\right) \tag{2}$$

The parameters a and b are replaced by 2^m and $2^m n$ correspondingly. Where 2^m and $2^m n$ are the dyadic scales. This kind of sampling eliminates redundancies in coefficients. Conceptually it is presented in Fig. 1.

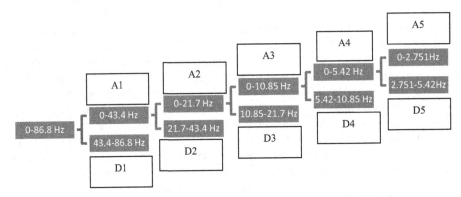

Fig. 1. Structure of five-level wavelet decomposition of EEG Signal

In DWT, high pass (HP) and low pass (LP) sub-bands are generated by filtering the signal using special HP and LP filters. The outputs obtained from the HP and LP filters are mentioned as an approximation (A) and details (D) coefficients. The outputs obtained from LP and HP filters at the first level are termed A1 and D1 correspondingly. For the next level, the same process filters the approximation coefficients (A1) and yield narrow sub-bands A2 & D2, and so on. With this, the signal of interest having large DWT coefficients can be captured, and the noisy signal with low DWT coefficients can be removed.

3.2 Discrete Wavelet Transforms

Pincus [35] first introduce the concept of approximate entropy (ApEn) to quantify the regularity and predictability of time series data in 1991. This parameter is very useful while extracting meaningful data from biological data. Regularity in EEG data can easily be achieved with the help of ApEn. Lower ApEn means the time series data have many redundancy patterns, while the higher ApEn refers to better unpredictability [36]. For a time-series data y_i having N sample points $y_1, y_2, y_3, \ldots\ldots, y_N$ the ApEn for fixed m, r, and τ in $\Re^m$ embedding space can be calculated as:

$$ApEn\ (m,\ r,\ \tau,\ N)\ =\ \phi^m(r) - \phi^{m+1}(r) \qquad (3)$$

where m = Embedding dimension,
r = Filtering level,
τ = time delay.

The $\phi^m(r)$ can be defined as:

$$\phi^m(r) = \frac{1}{N - (m-1)\tau} \sum_{i=1}^{N-(m-1)\tau} \log C_i^m(r) \tag{4}$$

For each i, correlation integral $(C_i^m(r))$ is as:

$$C_i^m(r) = \frac{1}{N - (m-1)\tau} \sum_{j=1} \Theta(r - \|x(i), x(j)\|) \tag{5}$$

Where,

$$\Theta(z) = \begin{cases} 1 \text{ for } z > 0 \\ 0 \text{ otherwise} \end{cases} \tag{6}$$

$$\|x(i), x(j)\| = \max_{t=1,2,3,\ldots,m} (|y(i + (t-1)\tau) - y(j + (t-1)\tau)|) \tag{7}$$

3.3 Hjorth Parameters

Hjorth parameters are characterized into three different categories, namely [37]:

1. Activity
2. Mobility
3. Complexity

The spectral properties of time-domain EEG signals can be extracted using the above parameters. Mathematically speaking.

1) The activity is the signal's variance, and it is high during seizure events.

$$\text{Activity} = var(y(t)) \tag{8}$$

Where $y(t)$ denotes the signal.
2) Mobility of the signal is given as

$$\text{Mobility} = \sqrt{\frac{var(\frac{dy(t)}{d(t)})}{var(y(t))}} \tag{9}$$

3) The complexity is known as

$$\text{Complexity} = \frac{\text{Mobility}\left(\frac{dy(t)}{dt}\right)}{\text{Mobility}y(t)} \tag{10}$$

3.4 Higuchi's Fractal Dimension (HFD)

In 1988, Higuchi derived a mathematical algorithm for finding the fractal dimension (FD) of discrete-time sequences [38]. This is a formula that calculates complexity of time series data. The value of HFD lies between 1 and 2, with a higher value referring to a higher level of signal complexity [39]. For a time-series data y_i having N sample points $y_1, y_2, y_3, \ldots \ldots, y_N$ the HFD defines k_{max} new time series as:

$$y_k^j : y_j, y_{(j+k)}, y_{(j+2k)}, \cdots \cdots \cdots, y_{(j+m)} \tag{11}$$

where $k_{max} > 1$, $m = int\left(\frac{N-j}{k}\right)k$, j is the first sample, $int\left(\frac{N-j}{k}\right)$ is the integer part of $\left(\frac{N-j}{k}\right)$, and $k = 1$ to k_{max}.

For each new time series y_k^j, the length can be calculated as:

$$L_k^j = \frac{1}{k}\left[\frac{N-1}{int\left(\frac{N-j}{k}\right)k}\left(\sum_{i=1}^{int\left(\frac{N-j}{k}\right)} \left|y_{(j+ik)} - y_{(j+(i-1)k)}\right|\right)\right] \tag{12}$$

The length of the curve L(k) for each possible k time step is obtained by averaging the k sets of L_k^j

$$L(k) = \frac{1}{k}\sum_{j=1}^{k} L_k^j \tag{13}$$

Finally, the curve is said to be FD if the plot of $\log(L(k))$ and $\log(1/k)$ lie on a straight line with slope $-FD$ i.e. $L(k) \sim k^{-FD}$.

3.5 Skewness and Kurtosis

Skewness and kurtosis are the 3rd and 4th order statistical moments. The skewness describes the data asymmetry, while the kurtosis measures the tailedness of data distribution. Specifically, both statistical moments indicate the shape of the distribution. These two statistical features have been used in developing an epileptic seizure classification model [17, 33], and achieved promising results. Mathematically, skewness is defined as:

$$Skew = \frac{E(X - \mu)^3}{\sigma^3} \tag{14}$$

While, kurtosis is defined as:

$$Kurt = \frac{E(X - \mu)^4}{\sigma^4} \tag{15}$$

Where, the expected value of any arbitrary variable X is represented as $E(X)$, the mean and the standard deviation of the signal is represented by μ and σ.

3.6 Support Vector Machines (SVMs)

Vapnik [40], in 1992, introduced a kernel-based machine-learning model for regression and classification tasks. Since their introduction, SVMs have been widely used for different classification tasks and have become one of the most used classifiers [41]. The reason for widespread adoption is the classifier's generalization capacity, low sensitivity to curse of dimensionality [42], and it is considered one of the best-known methods in terms of optimizing the solutions. SVMs are widely used for binary classification problems, and it has been shown that the performance is better than other supervised algorithms [41]. The objective of the machine learning models is to establish non-linear relationship between features and corresponding labels. SVMs do this by separating the data into two groups using a linear line called hyper-plane. Distance from this hyper-plane to the boundaries of two groups is called margin, and SVMs tries to maximize the margin to get better classification accuracy. However, most real-world data are not linearly separable. Therefore, with non-linear kernels (Gaussian, radial basis function, polynomial, etc.), the data points from non-linear input space are mapped to higher dimensional feature space which is linearly separable.

4 Materials and Methods

4.1 Datasets

4.1.1 Description of Dataset-1

The EEG data considered in this research was recorded at the University Hospital Bonn, Germany [43]. The dataset includes 5 subsets (represented as A, B, C, D and E). Total signals were recorded with the help of a 128-channel amplifier system. The subsets include 100 single-channel EEG segments of duration 23.6 s. The sampling frequency for the signal is 173.61 Hz. The EEG samples in subsets A and B are obtained from 5 healthy volunteers for both eyes open and closed conditions, with the help of external surface electrodes. The subsets C, D and E are the EEG segments of epileptic patients. The subsets C and D are captured in seizure-free intervals, and subset E consists of seizure activity. The whole data set can be classified into two categories. The one class that comprises epileptic seizure is composed of subset E, the other category, the non-seizure class, includes A, B, C and D. More information about the dataset can be found in [43].

4.1.2 Description of Dataset-2

The intended classification approach is also verified on other dataset to check the effectiveness of the proposed approach. Second dataset contains segmented EEG signals of ten (10) epileptic patients. The recordings were collected from Neurology and sleep center, Hauz Khas, New Delhi. The samples were collected using the Grass Telefactor Comet AS40 amplification system, having EEG electrodes placed as per the 10–20 placement system. In order to get the selective part of the frequency range (0.5–70 Hz), the signals were passed through the bandpass filter (BPF). The sampling rate is selected to be 200 Hz. The dataset was categorized and saved in three (3) folders named pre-ictal,

ictal, and inter-ictal, and each folder contains 50 MAT files of EEG segments. Each EEG segment comprises 1024 non-overlapping samples of the duration of 5.12 s. The pre-ictal data is labelled as Set X, inter-ictal as Set Y, and ictal as Set Z. Detailed description of the dataset is available in [29, 44].

4.2 Epileptic Seizure Detection

In this paper, an automated framework based on non-linear and statistical features and SVM for epileptic seizure detection is intended. First, the time domain signals are converted to time-frequency domain signals using DWT, and then the non-linear features such as ApEn, Higuchi's fractal dimension (HFrDm), Hjorth parameters (HjPm), statistical parameters (StPm) of D2, D3, D4, D5, and A5 sub-bands are calculated. Features calculated from these sub-bands combine into a feature vector. Then, the feature vector is normalized to avoid the domination of attributes having high numerical values over low numerical values during training and testing with SVMs classifier. Normalization of features can enhance the accuracy of the SVMs classifier [45]. When the feature vector is input to the SVMs classifier, then the classifier classifies the signal into epileptic and non-epileptic. Figure 2 shows the block diagram of the classification model. The authors have used the SVMs model with Gaussian, radial basis function and polynomial kernel and their performance is evaluated w.r.t sensitivity, specificity and accuracy.

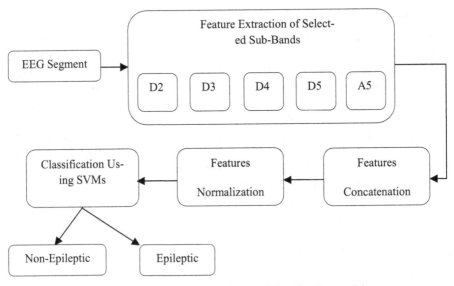

Fig. 2. Block diagram of the intended classification model

4.3 Performance Evaluation

Performance of the Classifier is computed w.r.t Sensitivity (Sens.), Specificity (Spec.), and Accuracy (Accu.) using the formula given in Eqs. 16, 17 and 18. The authors evaluated the performance by repeating 10-fold cross-validation 10 times each.

$$\text{Sensitivity} = \frac{TP}{TP + FN} \times 100\% \qquad (16)$$

$$\text{Specificity} = \frac{TN}{TN + FP} \times 100\% \qquad (17)$$

$$\text{Accuracy} = \frac{TP + TN}{TP + FN + TN + FP} \times 100\% \qquad (18)$$

where TP is True positive, TN is True negative, FP is false positive and FN is false negative.

5 Results and Discussion

SVMs model with different non-linear kernels (Gaussian, Radial Basis Function, and Polynomial) are tested on a complete set of features that includes the HFrDm, ApEn, HjPm, and StPm. The classifier's performance with different kernels (Gaussian, RBF, and Polynomial) is almost the same for all cases (A-E to ABCD-E, X-Z to XY-Z) as shown in Table 1. The comparison graph is also presented in Fig. 3.

Table 1. Comparison of results

Authors	Year	Features	Classifier	Case	Accuracy
N. Kannathal [21]	2005	Non-linear features	ANFIS	SetA against SetE	91.49%
V. Srinivasan [46]	2005	Time domain and Frequency domain features	Artificial Neural Network (ANN)	SetA against SetE	99.60%
V. Srinivasan [22]	2007	Non-linear features	ANN	SetA against SetE	100%
K. Polat [47]	2007	FFT	Decision Tree	SetA against SetE	98.72%
H. Ocak [14]	2009	DWT + ApEn	Statistical Model	SetACD against SetE	96%
U.R. Acharya [23]	2009	Non-linear features	Gaussian Mixture Model (GMM)	SetA against SetE	95%

(*continued*)

Table 1. (*continued*)

Authors	Year	Features	Classifier	Case	Accuracy
Guo [15]	2010	DWT + Line Length Features	ANN	SetA against SetE SetACD against SetE SetABCD against SetE	99.6% 97.75% 97.77%
Acharya [48]	2012	Non-linear features	Fuzzy	SetABCD against SetE	98.10%
Gandhi [30]	2012	Discrete Wavelet Packet Transform + Non-linear + Statistical features	Probabilistic Neural Network (PNN)	SetA against SetE	100%
V. Bajaj [33]	2012	Empirical Mode Decomposition (EMD)	LS-SVM	SetABCD against SetE	99.50–100%
Nicolaou [25]	2012		SVM	SetA against SetE SetB against SetE SetC against SetE SetD against SetE SetABCD against SetE	93.55% 82.88% 88.83% 83.13% 86.10%
M. Li [28]	2017	DWT based Envelop Analysis	Neural Network Ensemble	SetA against SetE	98.78%
S. Madan [17]	2017	DWT+Hurst Exponent	SVM	SetA against SetE SetC against SetE SetD against SetE SetBCD against SetE SetAB against SetE SetAC against SetE SetBC against SetE SetBD against SetE SetCD against SetE SetABC against SetE Set ACD against SetE SetABD against SetE SetABCD against SetE	99% 99% 93% 91.50% 94.67% 97.33% 93.33% 89.33% 94% 95.50% 95% 92.50% 93.20%
M. Mursalin [18]	2017	Time domain+ DWT+ Entropies	Random Forest	SetA against SetE SetB against SetE SetC against SetE SetD against SetE SetACD against SetE SetBCD against SetE SetCD against SetE SetABCD against SetE	100% 98% 99% 98.5% 98.5% 97.5% 98.67% 97.40%

(*continued*)

Table 1. (*continued*)

Authors	Year	Features	Classifier	Case	Accuracy
A. Sharmila [19]	2017	DWT+ ApEn+ Shannon Entropy	SVM	SetA against SetE	100%
				SetB against SetE	92.6%
				SetC against SetE	100%
				SetB against SetE	5.29%
				SetAC against SetE	81.94%
				SetAD against SetE	84.93%
				SetBC against SetE	82.43%
				SetBD against SetE	69.95%
				SetABC against SetE	94.17%
				SetABD against SetE	94.23%
				SetACD against SetE	79.54%
				SetBCD against SetE	69.44%
				SetABCD against SetE	82.19%
A.Subasi [49]	2017	Statistical Parameters of DWT Coefficients	Hybrid SVM	SetA against SetE	99.38%
A. Gupta [13]	2018		SVM	SetA against SetE	94.85%
				SetB against SetE	99%
				SetC against SetE	97.50%
				SetD against SetE	96.35%
				SetAB against SetE	97.27%
				SetCD against SetE	96.92%
				SetABCD against SetE	97.79%
				SetX against SetZ	79.70%
				SetY against SetZ	95.60%
				SetXY against SetZ	91.80%
Proposed Model	–	DWT+ Non-Linear Features	SVM	SetA against SetE	99.50%
				SetB against SetE	99.60%
				SetC against SetE	99.30%
				SetD against SetE	97.20%
				SetAB against SetE	99.33%
				SetAC against SetE	99.07%
				SetAD against SetE	97.63%
				SetBC against SetE	99.03%
				SetBD against SetE	97.67%
				SetCD against SetE	98.23%
				SetABC against SetE	99%
				SetABD against SetE	98.10%
				SetACD against SetE	98.15%
				SetBCD against SetE	98.32%
				SetABCD against SetE	98.50%
				SetX against SetZ	92.60%
				SetY against SetZ	100%
				SetXY against SetZ	96.53%

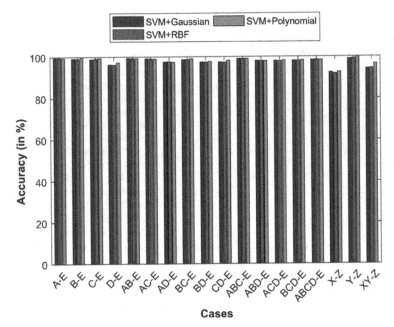

Fig. 3. Comparison graph of all classifiers for all cases

However, SVM with the polynomial kernel is the best performing model. The accuracy for most of the cases on dataset 1 is greater than 98%, while it is 97.20% and 97.67% for D against E and BD against E. Many cases like A against E, B against E, C against E, AB against E, AC against E, BC against E, and ABC against E have correctly classified with approximate 100% accuracy. For dataset 2, Y's classification accuracy against Z is around 100%, while it is 92.60% and 96.53% for X against Z and XY against Z. The accuracy drops for both the data sets (for pre-ictal vs ictal classification such as D against E and X against Z or with the inclusion of pre-ictal data in the same proportion with the normal data such as BD against E, XY against Z). However, the significance of qualitative features helps in classifying the seizure activity with the greatest possible accuracy. Thus, the features are robust and are almost independent of classifiers and types of kernels. The intended approach's accuracy is compared with the previous study reported on this dataset and is presented in Table 2.

Table 2: Overall performance of the classifiers (SVM + Gaussian, SVM + RBF, SVM + Polynomial) in % with respect to Sensitivity (Sens.), Specificity (Spec.), and Accuracy (Accu.) including all features (Fractal Dimension + Hjorth Parameters + Approximate Entropy + Statistical Parameters)

Case	Classifier								
	SVM + Gaussian			SVM + RBF			SVM + Polynomial		
	Sens.	Spec.	Accu.	Sens.	Spec.	Accu.	Sens.	Spec.	Accu.
A/E	100	99	99.50	100	99	99.50	100	98.60	99.30
B/E	99	99.20	99.10	99	99.10	99.05	100	99.20	99.60
C/E	97.60	100	98.80	98	100	99	100	98.60	99.30
D/E	96	96.50	96.25	96.10	96.20	96.15	96.50	97.90	97.20
AB/E	99.50	99	99.33	99.50	98.70	99.23	100	97.80	99.27
AC/E	99.10	99	99.07	99.05	98.90	99	99.85	97	98.90
AD/E	99.30	93.90	97.50	99.25	94.40	97.63	98.05	95.90	97.33
BC/E	98.45	98.90	98.60	98.55	98.90	98.67	99.70	97.70	99.03
BD/E	99.25	93.70	97.40	99.20	93.60	97.33	98.15	96.70	97.67
CD/E	97.85	96.60	97.43	97.75	96.30	97.27	98.30	98.10	98.23
ABC/E	99.07	98.70	98.98	99.10	98.50	98.95	99.80	96.60	99
ABD/E	99.60	93.60	98.10	99.50	93.40	97.97	98.83	95.30	97.95
ACD/E	99.47	93.70	98.02	99.33	93.50	97.88	98.97	95.70	98.15
BCD/E	99.50	93.70	98.05	99.57	93.60	98.07	98.83	96.80	98.32
ABCD/E	99.65	93.70	98.46	99.75	93.50	98.50	99.10	95.60	98.40
X/Z	93.80	91	92.40	94	89.60	91.80	91.60	93.60	92.60
Y/Z	99.20	98.60	98.90	99.80	98.40	99.10	100	100	100
XY/Z	97.20	88.20	94.20	97.30	88.60	94.40	97.90	93.80	96.53

6 Conclusion

Feature selection plays an important part in improvising the performance of any Model based on Machine learning. The qualitative nature of the feature selected for this intended model has enhanced the performance with respect to accuracy, sensitivity and specificity. The experimental results shows qualitative nature of multiple non-linear and statistical features considered in this research, as the performance of the intended model is better in comparison to the previous reported results. The classifier's performance with different non-linear kernels is almost the same for all cases (A- E to ABCD- E, X- Z to XY- Z). The reason for this performance is due to the consideration of multiple non-linear features. Different non-linear measures capture different information, so combining multiple non-linear features minimises the risk of losing the necessary information and contributes to better seizure classification accuracy. Since the proposed approach provides a computationally inexpensive solution, therefore the intended approach may be used in clinical practices.

References

1. World Health Organization: Global status report on alcohol and health 2018. World Health Organization (2019)
2. Cook, M.J., et al.: Prediction of seizure likelihood with a long-term, implanted seizure advisory system in patients with drug-resistant epilepsy: a first-in-man study. Lancet Neurol. **12**(6), 563–571 (2013)
3. Moghim, Negin, and David W. Corne. "Predicting epileptic seizures in advance." *PloS one* 9, no. 6 (2014): e99334.
4. Santhosh, N.S., Sinha, S., Satishchandra, P.: Epilepsy: Indian perspective. Ann. Indian Acad. Neurol. **17**(Suppl 1), S3 (2014)
5. Meinardi, H., Scott, R.A., Reis, R., Sander, J.W.A.S.: On behalf of the ılae commission on the developing world. The treatment gap in epilepsy: the current situation and ways forward. Epilepsia **42**(1), 136–149 (2001)
6. Scott, R.A., Lhatoo, S.D., Sander, J.W.A.S.: The treatment of epilepsy in developing countries: where do we go from here? Bull. World Health Organ. **79**, 344–351 (2001)
7. Leonardi, M., Ustun, T.B.: The global burden of epilepsy. Epilepsia **43**, 21–25 (2002)
8. Gourie-Devi, M.: Epidemiology of neurological disorders in India: review of background, prevalence and incidence of epilepsy, stroke, Parkinson's disease and tremors. Neurol. India **62**(6), 588 (2014)
9. Oto, M.M.: The misdiagnosis of epilepsy: appraising risks and managing uncertainty. Seizure **44**, 143–146 (2017)
10. Panayiotopoulos, C.P.: Epileptic Syndromes and Their Treatment. Neonatal Seizures. 2nd ed. London, pp. 185–206 (2007)
11. Mei, Z., Zhao, X., Chen, H., Chen, W.: Bio-signal complexity analysis in epileptic seizure monitoring: a topic review. Sensors **18**(6), 1720 (2018)
12. Chen, H., Koubeissi, M.Z.: Electroencephalography in epilepsy evaluation. CONTINUUM Lifelong Learn. Neurol. **25**(2), 431–453 (2019)
13. Gupta, A., Singh, P., Karlekar, M.: A novel signal modeling approach for classification of seizure and seizure-free EEG signals. IEEE Trans. Neural Syst. Rehabil. Eng. **26**(5), 925–935 (2018)
14. Ocak, H.: Automatic detection of epileptic seizures in EEG using discrete wavelet transform and approximate entropy. Expert Syst. Appl. **36**(2), 2027–2036 (2009)
15. Guo, L., Rivero, D., Dorado, J., Rabunal, J.R., Pazos, A.: Automatic epileptic seizure detection in EEGs based on line length feature and artificial neural networks. J. Neurosci. Methods **191**(1), 101–109 (2010)
16. Chen, L.-L., Zhang, J., Zou, J.-Z., Zhao, C.-J., Wang, G.-S.: A framework on wavelet-based nonlinear features and extreme learning machine for epileptic seizure detection. Biomed. Signal Process. Control **10**, 1–10 (2014)
17. Madan, S., Srivastava, K., Sharmila, A., Mahalakshmi, P.: A case study on discrete wavelet transform based hurst exponent for epilepsy detection. J. Med. Eng. Technol. **42**(1), 9–17 (2018)
18. Mursalin, M., Zhang, Y., Chen, Y., Chawla, N.V.: Automated epileptic seizure detection using improved correlation-based feature selection with random forest classifier. Neurocomputing **241**, 204–214 (2017)

19. Sharmila, A., Raj, S.A., Shashank, P., Mahalakshmi, P.: Epileptic seizure detection using DWT-based approximate entropy, Shannon entropy and support vector machine: a case study. J. Med. Eng. Technol. **42**(1), 1–8 (2018)
20. Chen, D., Wan, S., Xiang, J., Bao, F.S.: A high-performance seizure detection algorithm based on Discrete Wavelet Transform (DWT) and EEG. PloS One **12**(3), e0173138 (2017)
21. Kannathal, N., Choo, M.L., Rajendra Acharya, U., Sadasivan, P.K.: Entropies for detection of epilepsy in EEG. Comput. Methods Programs Biomed. **80**(3), 187–194 (2005)
22. Srinivasan, V., Eswaran, C., Sriraam, N.: Approximate entropy-based epileptic EEG detection using artificial neural networks. IEEE Trans. Inf. Technol. Biomed. **11**(3), 288–295 (2007)
23. Acharya, U.R., Chua, C.K., Lim, T.-C., Dorithy, Suri, J.S.: Automatic identification of epileptic EEG signals using nonlinear parameters. J. Mech. Med. Biol. **9**(04), 539–553 (2009)
24. Kang, J.-H., Chung, Y.G., Kim, S.-P.: An efficient detection of epileptic seizure by differentiation and spectral analysis of electroencephalograms. Comput. Biol. Med. **66**, 352–356 (2015)
25. Nicolaou, N., Georgiou, J.: Detection of epileptic electroencephalogram based on permutation entropy and support vector machines. Expert Syst. Appl. **39**(1), 202–209 (2012)
26. Kumar, Y., Dewal, M.L., Anand, R.S.: Epileptic seizures detection in EEG using DWT-based ApEn and artificial neural network. SIViP **8**(7), 1323–1334 (2012). https://doi.org/10.1007/s11760-012-0362-9
27. Sharma, M., Pachori, R.B., Rajendra Acharya, U.: A new approach to characterize epileptic seizures using analytic time-frequency flexible wavelet transform and fractal dimension. Pattern Recogn. Lett. **94**, 172–179 (2017)
28. Li, M., Chen, W., Zhang, T.: Classification of epilepsy EEG signals using DWT-based envelope analysis and neural network ensemble. Biomed. Signal Process. Control **31**, 357–365 (2017)
29. Hussain, L.: Detecting epileptic seizure with different feature extracting strategies using robust machine learning classification techniques by applying advance parameter optimization approach. Cogn. Neurodyn. **12**(3), 271–294 (2018). https://doi.org/10.1007/s11571-018-9477-1
30. Gandhi, T.K., Chakraborty, P., Roy, G.G., Panigrahi, B.K.: Discrete harmony search based expert model for epileptic seizure detection in electroencephalography. Expert Syst. Appl. **39**(4), 4055–4062 (2012)
31. Burns, T., Rajan, R.: Combining complexity measures of EEG data: multiplying measures reveal previously hidden information. F1000Research **4** (2015)
32. Liang, Z., et al.: EEG entropy measures in anesthesia. Front. Comput. Neurosci. **9**, 16 (2015)
33. Bajaj, V., Pachori, R.B.: Classification of seizure and nonseizure EEG signals using empirical mode decomposition. IEEE Trans. Inf. Technol. Biomed. **16**(6), 1135–1142 (2011)
34. Anand, S., Jaiswal, S., Ghosh, P.K.: Automatic focal epileptic seizure detection in EEG signals. In: 2017 IEEE International WIE Conference on Electrical and Computer Engineering (WIECON-ECE), pp. 103–107. IEEE (2017)
35. Pincus, S.M.: Approximate entropy as a measure of system complexity. Proc. Natl. Acad. Sci. **88**(6), 2297–2301 (1991)
36. Tsafack, N., et al.: A new chaotic map with dynamic analysis and encryption application in internet of health things. IEEE Access **8**, 137731–137744 (2020)
37. Hjorth, B.: EEG analysis based on time domain properties. Electroencephalogr. Clin. Neurophysiol. **29**(3), 306–310 (1970)
38. Higuchi, T.: Approach to an irregular time series on the basis of the fractal theory. Physica D **31**(2), 277–283 (1988)
39. Smits, F.M., Porcaro, C., Cottone, C., Cancelli, A., Rossini, P.M., Tecchio, F.: Electroencephalographic fractal dimension in healthy ageing and Alzheimer's disease. PloS One **11**(2), e0149587 (2016)

40. Boser, B.E., Guyon, I.M., Vapnik, V.N.: A training algorithm for optimal margin classifiers. In: Proceedings of the Fifth Annual Workshop on Computational Learning Theory, pp. 144–152 (1992)

41. Cervantes, J., Garcia-Lamont, F., Rodríguez-Mazahua, L., Lopez, A.: A comprehensive survey on support vector machine classification: applications, challenges and trends. Neurocomputing **408**, 189–215 (2020)

42. Khan, M.H., Saleem, Z., Ahmad, M., Sohaib, A., Ayaz, H., Mazzara, M.: Hyperspectral imaging for color adulteration detection in red chili. Appl. Sci. **10**(17), 5955 (2020)

43. Andrzejak, R.G., Lehnertz, K., Mormann, F., Rieke, C., David, P., Elger, C.E.: Indications of nonlinear deterministic and finite-dimensional structures in time series of brain electrical activity: dependence on recording region and brain state. Phys. Rev. E **64**(6), 061907 (2001)

44. Swami, P., Gandhi, T.K., Panigrahi, B.K., Tripathi, M., Anand, S.: A novel robust diagnostic model to detect seizures in electroencephalography. Expert Syst. Appl. **56**, 116–130 (2016)

45. Hsu, C.W., Chang, C.C., Lin, C.J.: A practical guide to support vector classification. National Taiwan University, Taiwan, Technical report (2010)

46. Srinivasan, V., Eswaran, C., Sriraam, N.: Artificial neural network based epileptic detection using time-domain and frequency-domain features. J. Med. Syst. **29**(6), 647–660 (2005)

47. Polat, K., Güneş, S.: Classification of epileptiform EEG using a hybrid system based on decision tree classifier and fast Fourier transform. Appl. Math. Comput. **187**(2), 1017–1026 (2007)

48. Acharya, U.R., Molinari, F., Sree, S.V., Chattopadhyay, S., Ng, K.-H., Suri, J.S.: Automated diagnosis of epileptic EEG using entropies. Biomed. Sig. Process. Control **7**(4), 401–408 (2012)

49. Subasi, A., Kevric, J., Abdullah Canbaz, M.: Epileptic seizure detection using hybrid machine learning methods. Neural Comput. Appl. **31**(1), 317–325 (2017). https://doi.org/10.1007/s00521-017-3003-y

Machine Learning Applications in Smart Cities

Multimodal Cyberbullying Detection Using Ensemble Learning

Piyush Khanna, Abhinav Mathur, Anunay Chandra, and Akshi Kumar[✉]

Department of Computer Science and Engineering, Delhi Technological University,
Delhi, India
{piyushkhanna_bt2k17,abhinavmathur_2k17co08,
anunaychandra_2k17co71}@dtu.ac.in, akshikumar@dce.ac.in

Abstract. Modern technologies have made the internet more accessible, with social networks and online communications witnessing a several-fold increase in usage and popularity. Since it offers such ease and convenience, it has become an indispensable part of modern culture. However, as with all good things, it has also led to a new peril - Cyberbullying, which is essentially bullying an individual via an electronic medium. With a high amount of data being shared on social networks daily, automated cyberbullying detection tools need to be put in place. Several approaches have been made for text-based cyberbullying detection, but other modalities such as images have often been ignored. In this paper, we propose a multimodal Long Short-Term Memory (LSTM) network for cyberbullying detection that captures the interplay between the textual and visual modalities by conditioning the LSTM on nontemporal visual data. We further use this network to create an ensemble model that achieves an F1 score of 0.81, outperforming the current state-of-the-art model by 3.8% on the same dataset.

Keywords: Cyberbullying detection · Online social network · Ensemble learning

1 Introduction

Significant advances in technology have made the world smaller by bringing us all together. The internet has revolutionized communication, changing the way we create knowledge, consume content, and communicate. The influence of digital content on people, especially young adults, has hit unprecedented highs [1]. Since there are no space and time barriers, the internet has opened up a wide range of communicative possibilities. However, as with all technologies, some people have misused their power for a social peril-Cyberbullying.

Cyberbullying is the inappropriate usage of technology such as the internet, cell phones, or other electronic devices to bully someone [2]. Victims are often targeted through SMSes, social networks, and online forums. It includes posting, sending, or sharing negative content that is harmful, mean, or obscene. The

© Springer Nature Switzerland AG 2021
A. Solanki et al. (Eds.): AIS2C2 2021, CCIS 1434, pp. 221–229, 2021.
https://doi.org/10.1007/978-3-030-82322-1_16

biggest concern regarding cyberbullying is the anonymity offered by the internet. Shielded by anonymity, individuals don't need to take responsibility for their actions, because of which they tend to be more direct and intense [3]. Cyberbullying victims tend to suffer from long-lasting effects such as depression, stress, and anxiety, among others [4]. Given the seriousness of the situation, cyberbullying detection tools have become the need of the hour. Due to the high volume of data being shared over the internet, [5], automated detection tools are required.

Several approaches have been made to detect cyberbullying in texts from posts, tweets, and messages. However, most social networks now use images and videos with Text as well. Image sharing sites such as Instagram and Snapchat are becoming increasingly popular among teenagers [6]. With an increase in cyberbullying through images and video being reported [7], tools for cyberbullying detection in multiple modalities need to be implemented. In this paper, we propose a multimodal ensemble model for cyberbullying detection on social media posts containing Text and corresponding images. Our approach utilizes a multimodal LSTM network for leveraging the interplay between the textual-visual modalities. We further use this network to create an ensemble model to get better and robust performance.

2   Related Work

Historically, cyberbullying detection has focused on the textual modality, ignoring other factors. Recently, however, there has been a growing interest in studying multimodal data sources to identify instances of cyberbullying in Online Social Networks. We classify the relevant work as i) Text-based detection ii) Text and image-based detection.

2.1   Text Based Cyberbullying Detection

Prior work in text-based detection has employed simple classifiers such as SVM, Random Forest and K-Nearest Neighbours Classifier on Twitter posts to detect cyberbullying [8–11]. Recent work also focuses on meta-data attributes to identify such instances such as gender [12] and sexual orientation [13] of victims, bullying based on racism and sexism [14] and other soft computing techniques [15] for detecting bullying on social multimedia. There has also been a significant focus on analysing code-mixed data for cyberbullying detection [16]. However, these techniques consider only unimodal data sources to detect cyberbullying, whereas most of the data shared on social networks is multimodal in nature.

2.2   Text and Image Based Detection

With the increase of multiple modalities in online social networks, cyberbullying detection using multimodal data has become a critical area of research. Recent work in this domain employed capsule network along with convolutional networks to study multimodal cyberbullying detection [17].[18] considered user data such as follower count along with visual features like image and caption. According to their findings, visual features weren't helpful. [19] used a small training sample containing highly negative words, along with visual features. [20] collected a dataset of 2100 images from Twitter, Facebook and Instagram. Comments were manually added for some images. They used TF-IDF representation for textual features, and convolutional neural network for visual features, achieving F1-score of 0.68. [21] used CNN and VGG-16 models to process embedded text and images respectively via a genetic algorithm. Although, inclusion of multiple modalities enhances detection, prior work only considers simple fusion of such modalities. To address this, we propose an image-aware textual multimodal LSTM model to fuse the modalities effectively.

3   Methodology

3.1   Text Processing

We represent the textual component of each post using 300-dimensional Glove-6B [22] embedding. We zero-pad the embedding that has less than the maximum number of words for efficient batching. We further employ a multimodal LSTM network as described in Sect. 3.3 to extract multimodal features from the textual embedding combined with the visual embedding of the corresponding image.

3.2   Image Processing

Deep neural networks have demonstrated remarkable performance in image recognition in recent times [23,24]. Inception-v3 [24] is a state-of-the-art model optimized to perform well in constrained memory requirements and computational budget. In our approach, we use the TensorFlow implementation[1] of the Inception-v3 model pre-trained on the ImageNet dataset as a feature extraction module. These features are further flattened and fed to a dense layer to obtain image embeddings used in our multimodal models.

3.3   Multimodal LSTM for Cyberbullying Detection

Prior work in the cyberbullying context [20,21] has demonstrated the advantage of using multimodal data for identifying online bullying. However, identifying such bullying on social media poses a challenge due to the heterogeneous multimodal form of content. The temporal nature of multiple modalities, such as Text, which requires the context of previous words in a sentence for a meaningful representation, and image, which has no such constraints for a static image, may not be consistent with each other, making this task complicated. To overcome this disparity, we employ multimodal "conditioned" LSTM network [25]. This network provides a reasonable approach to fuse nontemporal data (image) with temporal data (Text).

---

[1] InceptionNet implementation.

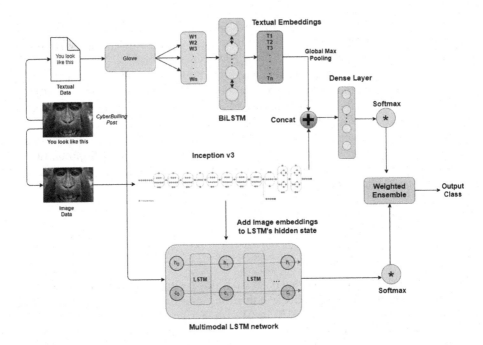

Fig. 1. Schematic diagram for model architecture.

In our case, as shown in Fig. 1, we pass the Glove embedding through an LSTM network and condition the LSTM on nontemporal visual data by adding the image embedding to its hidden state at the first timestep. Formally, training the LSTM model takes the sequence input vectors $(x_1, .., x_T)$ representing the Glove embedding of the Text along with the image embedding $\mathcal{I}$ extracted using Inception for each post. The model computes a series of hidden states $(h_1, .., h_T)$ and a sequence of outputs $(y_1, ..y_T)$, by repeating the following recurrence relation from time t = 1 to T:

$$h_t = f(W_{hx}x_t + W_{hh}h_{t-1} + \mathcal{I} + b_h) \tag{1}$$

$$y_T = softmax(W_{oh}h_t + b_o) \tag{2}$$

Here, $W_{hx}, W_{hh}, W_{oh}, x_i, b_h, b_o$ are learnable parameters. We use the output y_T from the last LSTM unit and pass it through softmax activation for the final prediction output.

3.4 Multimodal Concatenation Model

For our second model in the ensemble approach, we perform a simple concatenation between the embeddings of the two modalities. In this case, we use a BiLSTM model to encode the textual features. The textual features are then max-pooled to make their shape consistent with the image embeddings obtained using Inception for concatenation.

Formally, we represent the words in the Text of each post as $(w_1, w_2, ..., w_n)$ where w_i is the ith word and n is the number of words which are encoded as follows:

$$\overrightarrow{T_t^{(f)}} = BiLSTM^{(f)}(w_t, T_{t-1}^{(f)}) \qquad (3)$$

$$\overleftarrow{T_t^{(b)}} = BiLSTM^{(b)}(w_t, T_{t+1}^{(b)}) \qquad (4)$$

$$T_t = [\overrightarrow{T_t^{(f)}}, \overleftarrow{T_{T-t}^{(b)}}] \qquad (5)$$

We perform a global max-pooling operation on the textual 1D temporal output obtained from the last timestep of the BiLSTM network T_n to make it compatible for concatenation with the 1D image embeddings $\mathcal{I}$. The visual-textual concatenated features are then fed to a dense layer with softmax activation for the final prediction output.

$$y = softmax(\text{concat}(\mathcal{I}, \text{GlobalMaxPool}(T_n))) \qquad (6)$$

3.5   Ensemble Approach

Finally, as shown in Fig. 1, we ensemble the above described models in a late-fusion fashion by computing a weighted average of the predicted class probabilities. If $p_{Multi-LSTM}$ and $p_{Multi-Concat}$ represent the class probabilities of the Multimodal LSTM model and the Multimodal Concatenation model respectively and α is a variable ranging from 0 to 1, then the final ensemble class probability $p_{ensemble}$ is given by:

$$p_{ensemble} = \alpha * p_{Multi-Concat} + (1 - \alpha) * p_{Multi-LSTM} \qquad (7)$$

4   Dataset

We used the cyberbullying dataset [20] collected from various social networks such as Twitter, Instagram and Facebook for all experiments. Figure 2 shows sample illustrations from the dataset. The dataset consists of 2100 social media posts(619 non-bullying and 1481 bullying). Each post comprises of an image along with a comment. Following [20], we randomly pick 619 out of 1481 samples from the bullying class in order to balance the dataset and have equal number of samples from both classes. Finally, we divide the data into train, validation and test sets in the ratio 65:10:25, respectively. We compare our approach with recent state-of-the-art models against F1, Precision and Recall.

IMAGES				
TEXT	What are you looking up?	You are beautiful	Policemen always bark.	Come to Gym we have planned something?
LABEL	Non-Bullying	Non-Bullying	Bullying	Bullying

Fig. 2. Images and texts from the dataset.

5   Experiment Setting

We tune our model hyperparameters based on the following search space on the validation set: hidden layer size, dropout $\in [0, 0.7]$, batch size $\in \{16, 32, 64\}$, epochs ($<$100), learning rate $\in \{10^{-3}, 10^{-2}, 10^{-1}\}$, ensemble ratio $\alpha \in [0, 1]$ varied with step-size$= 0.1$. We report the average performance of our model by randomly running 10 splits on our dataset.

Baselines: We present a comparison between our approach and the baseline models in Table 1. Following [21], we use 1-D CNN and LSTM models for our unimodal text-based models and CNN, VGG-16 and Inception for unimodal image-based models. We also compare our approach against current state-of-the-art models such as 2-D Unified Representation model [20] which represents both textual and visual data in the form of 2D matrices which are later fused and processed using a CNN, and Multimodal model using Genetic Algorithm [21], which uses a genetic algorithm to optimize the multimodal feature set obtained using VGG-16 (for images) and CNN (for Text).

6   Results

6.1   Comparative Analysis

Table 1 shows the quantitative comparisons against the baseline models. We observe that our Multimodal-LSTM approach has similar performance compared to the current state-of-the-art Multimodal Genetic Algorithm model [21] and outperforms the 2-D Unified Representation model [20]. This performance could be attributed to a better representation of textual features, as it preserves the word ordering and context of the text while blending it with the corresponding image's context in the LSTM unit itself. Hence, the Multimodal-LSTM model is a novel approach for cyberbullying detection that fuses the inter-modal relations by conditioning the textual features based on the input image's content. Furthermore, our weighted average ensemble approach shows substantial improvement compared to any single model, as the ensemble assigns appropriate weightage to the models based on their respective contribution. Figure 3b shows the variation of F1 with $(1 - \alpha)$ on the validation set as illustrated by Eq. 7. We found the best $1 - \alpha = 0.3$ $(\alpha = 0.7)$.

Table 1. Weighted Precision, Recall and F1 scores of baseline approaches.

Data	Approach	Precision	Recall	F1
Text	CNN	0.74	0.75	0.75
	LSTM	0.64	0.65	0.64
Image	CNN	0.66	0.65	0.66
	VGG-16	0.7	0.7	0.7
Multimodal	2-D Unified Representation (2D UR) [20]	0.69	0.69	0.69
	Multimodal Genetic Algorithm (MGA) [21]	0.80	0.79	0.78
	Multimodal-Concat (MC)	0.72	0.76	0.75
	Multimodal-LSTM (MLSTM)	0.80	0.77	0.78
	Our Ensemble Approach	**0.82**	**0.81**	**0.81**

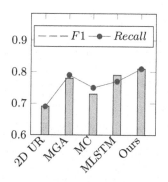

(a) Comparative analysis

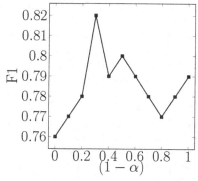

(b) $(1 - \alpha)$ vs F1 on validation set.

Fig. 3. Quantitative Analysis of our approach.

6.2 Conclusion and Future Work

Cyberbullying is a major threat to the mental and emotional well-being of millennials. Social media posts provide a rich source of multimodal data that can be used to study and detect cyberbullying. However, this area remains underexplored. To this end, we propose an ensemble learning-based approach that outperforms the current state-of-the-art models using multimodal LSTM network that better captures the interplay between the textual and visual modalities. Our future work involves studying the network effects on a social media platform and how they influence cyberbullying.

References

1. Adhiarso, D., Utari, P., Hastjarjo, S.: The impact of digital technology to change people's behavior in using the media. **2**, 06 (2019)
2. Smith, P., Mahdavi, J., Carvalho, M., Fisher, S., Russell, S., Tippett, N.: Cyberbullying: its nature and impact in secondary school pupils. J. Child Psychol. Psychiatry Allied Disciplines **49**, 376–385 (2008)
3. Suler, J.: The online disinhibition effect. Cyberpsychol. Behav. Impact Multimedia Virtual Reality Behav. Soc. **7**, 321–326 (2004)
4. Nixon, C.: Current perspectives: the impact of cyberbullying on adolescent health. Adolesc. Health Med. Ther. **5**, 143–158 (2014)
5. Devakunchari, R.: Analysis on big data over the years. Int. J. Sci. Res. Publ. **4**(1), 1–7 (2014)
6. Pater, J., Miller, A., Mynatt, E.: This digital life: a neighborhood-based study of adolescents' lives online (2015)
7. Seiler, S.J., Navarro, J.N.: Bullying on the pixel playground: investigating risk factors of cyberbullying at the intersection of children's online-offline social lives. Cyberpsychol. J. Psychosoc. Res. Cyberspace 8(4)
8. Al-garadi, M.A., Varathan, K.D., Ravana, S.D.: Cybercrime detection in online communications: the experimental case of cyberbullying detection in the twitter network. Comput. Hum. Behav. **63**, 433–443 (2016)
9. Yin, D., Xue, Z., Hong, L., Davison, B., Edwards, A., Edwards, L.: Detection of harassment on web 2.0. (2009)
10. Reynolds, K., Edwards, A., Edwards, L.: Using machine learning to detect cyberbullying. In: Proceedings of 10th International Conference on Machine Learning and Applications, ICMLA 2011. vol. 2, p. 12 (2011)
11. Burnap, P., Williams, M.: Cyber hate speech on twitter: an application of machine classification and statistical modeling for policy and decision making: machine classification of cyber hate speech. Policy & Internet **7**, 223–242 (2015)
12. Dadvar, M., De Jong, F.: Cyberbullying detection; a step toward a safer internet yard (2012)
13. Dinakar, K., Jones, B., Havasi, C., Lieberman, H., Picard, R.: Common sense reasoning for detection, prevention, and mitigation of cyberbullying. ACM Trans. Interact. Intell. Syst. **2**, 1–30 (2012)
14. Chen, H., Mckeever, S., Delany, S.J.: Harnessing the power of text mining for the detection of abusive content in social media. In: Angelov, P., Gegov, A., Jayne, C., Shen, Q. (eds.) Advances in Computational Intelligence Systems. AISC, vol. 513, pp. 187–205. Springer, Cham (2017). https://doi.org/10.1007/978-3-319-46562-3_12
15. Kumar, A., Sachdeva, N.: Cyberbullying detection on social multimedia using soft computing techniques: a meta-analysis. Multimedia Tools Appl. **78**(17), 23973–24010 (2019). https://doi.org/10.1007/s11042-019-7234-z
16. Kumar, A., Sachdeva, N.: Multi-input integrative learning using deep neural networks and transfer learning for cyberbullying detection in real-time code-mix data. Multimedia Syst. 1–15 (2020). https://doi.org/10.1007/s00530-020-00672-7

17. Kumar, A., Sachdeva, N.: Multimodal cyberbullying detection using capsule network with dynamic routing and deep convolutional neural network. Multimedia Syst. 1–10 (2021). https://doi.org/10.1007/s00530-020-00747-5
18. Hosseinmardi, H., Rafiq, R.I., Han, R., Lv, Q., Mishra, S.: Prediction of cyberbullying incidents in a media-based social network. pp. 186–192 (2016)
19. Singh, V.K., Ghosh, S., Jose, C.: Toward multimodal cyberbullying detection. In: CHI 2017 Extended Abstracts - Proceedings of the 2017 ACM SIGCHI Conference on Human Factors in Computing Systems, Conference on Human Factors in Computing Systems - Proceedings, pp. 2090–2099. Association for Computing Machinery, May 2017. In: 2017 ACM SIGCHI Conference on Human Factors in Computing Systems, CHI EA 2017, Conference date: 06–05-2017 Through 11–05-2017
20. Kumari, K., Singh, J.P., Dwivedi, Y.K., Rana, N.P.: Towards cyberbullying-free social media in smart cities: a unified multi-modal approach. Soft Comput. 24(15), 11059–11070 (2020)
21. Kumari, K., Singh, J.P.: Identification of cyberbullying on multi-modal social media posts using genetic algorithm. Trans. Emerg. Telecommun. Technol. 32(2), e3907 (2020)
22. Pennington, J., Socher, R., Manning, C.D.: Glove: global vectors for word representation. In: Empirical Methods in Natural Language Processing (EMNLP), pp. 1532–1543 (2014)
23. Szegedy, C.: Going deeper with convolutions. In: Proceedings of the IEEE Conference on Computer Vision and Pattern Recognition, pp. 1–9 (2015)
24. Szegedy, C., Vanhoucke, V., Ioffe, S., Shlens, J., Wojna, Z.: Rethinking the inception architecture for computer vision. In: Proceedings of the IEEE Conference on Computer Vision and Pattern Recognition, pp. 2818–2826 (2016)
25. Karpathy, A., Fei-Fei, L.: Deep visual-semantic alignments for generating image descriptions. In: Proceedings of the IEEE Conference on Computer Vision and Pattern Recognition, pp. 3128–3137 (2015)

Frequent Route Pattern Mining Technique for Route Prediction in Transportation Network

Pritam[✉] [iD], Dimple Singh[✉] [iD], Komal Kumar Bhatia[✉] [iD],
and Neelam Duhan[✉] [iD]

J C Bose University of Science and Technology, YMCA, Faridabad, India

Abstract. People movement pattern plays a vital role in many applications like route prediction, route planning, traffic management, location-aware computing, traffic congestion estimation and so on. However, understanding movement pattern is quite challenging such as filtering GPS coordinates, mining of periodic pattern, finding credible and reliable information from the huge dataset and so on. The process of finding unexpected, frequent, and useful patterns from the data is called pattern mining. This paper proposes a novel frequent pattern mining algorithm for understanding the movement patterns of people. Discovering such movement patterns from trajectory data helps in finding periodic behavior. Further, these patterns can be used for predicting, planning, and managing routes. The experiment is performed using real datasets which contains both spatial and non-spatial datasets. These datasets are open source.

Keywords: Pattern mining · Route pattern mining · Route prediction · GPS · Trajectory

1 Introduction

Pattern Mining is the process of finding useful patterns from database. It is a key requirement of many scientific & commercial applications such as protein comparison [12], construction of sequential classification models, document comparison [13] and so on. There are various pattern-mining approaches like frequent pattern mining, frequent pattern mining with uncertain data, frequent graph-based substructure pattern mining, open-source pattern mining, sequential pattern mining, Constraint-based sequential pattern mining, structured pattern mining and so on.

The Trajectory is represented by a sequence of geographical locations with timestamps. These geographical locations with time stamps are termed as GPS-coordinates. These coordinates can be obtained from Mobile phones with GPS capabilities or GPS-enabled devices. The collection of these GPS coordinates is used to make a sequential database. In route pattern mining, a sequential database is a collection of trajectories of people [10].

The main objective of the proposed method is to find the frequent patterns from the sequential database. These patterns can be used in route prediction domain. Route

A. Solanki et al. (Eds.): AIS2C2 2021, CCIS 1434, pp. 230–241, 2021.
https://doi.org/10.1007/978-3-030-82322-1_17

Prediction is the basic requirement of intelligent transport system (ITS) services such as travel pattern similarity, traffic congestion estimation, Eco-routing [11] and so on.

The main contribution of this study is the application of route pattern mining in Route Prediction. This will significantly improve the existing transportation system services and improve ITS services. Many applications in the field of Intelligent Transportation system services have motivated the researchers. The researchers are trying to explore the possibilities of Route Pattern Mining [14].

The rest of the paper is organized as follows: Sect. 2 gives a detailed literature review of some recent research in pattern mining, Sect. 3 proposes a novel approach, Sect. 4 presents the implementation of the proposed pattern mining approach. In addition, the experimental results are also presented in it. Finally, Sect. 5 concludes the paper.

Definition 1: A *Trajectory* $t_{1:k}$ is a collection of k GPS data points and each GPS data point is a 3-tuple (t_k.*latitude*, t_k.*longitude*, t_k.*time*) indicates latitude, longitude and time stamp respectively [18].

2 Related Work

The study is related to pattern mining in different domain. There are several approaches to tackle pattern mining problem such as substring tree structure and improved level-wise mining algorithm which is proposed by Cao et al. [6], MiSTA algorithm which is proposed by Giannotti et al. [7], projection- based algorithm which is proposed by Gidofalvi et al. [8], Combination of K-means clustering and FP-Growth algorithm which is suggested by Sandhya et al. [9] for effective data partitioning and so on.

Frederic Stahl et al. [1] proposed a novel predictive algorithm i.e. Fast Generalized Rule Induction (FGRI). It induces descriptive rulesets for streaming data. While inducing rules, only the most frequent feature-value pairs are considered. This speed up the rule induction process. It also depicts the accuracy of descriptive rules. It is especially designed for handling unlabeled real-time streaming data. It is responsive as well as adaptive to new data instances. This implies that rules are expressive, expandable, and applicable on new data streams. Both categorical and numerical features are allowed in data stream. The method focuses on parts of the rule set that are affected to generate new rulesets. There is a no need of complete retraining whenever a new data stream came to generate updates rulesets.

Ming-Yen Lin et al. [2] proposed a novel algorithm, Quantitative high-utility item-sets mining (QHIM). The item-sets with high profit is generated using this algorithm. These item-sets are called as high utility item-sets. Discovering high utility item-sets is the main objective of this paper. A pattern-growth-based structure is maintained for storing this quantitative information. It is a two-step process. It involves the construction of global header table and global quantitative utility tree (QU-tree). Then the next step is discovering of all item-sets of QHIM. A search space is needed to store QU-tree and header table. The generation of candidate item-sets require two database scans only. The algorithm performs well for small data. The experimental results show that it is an efficient algorithm when the minimum threshold is low.

M. Sornalakshmi et al. [3] projected the enhanced parallel and distributed a-priori (EPDA) algorithm for healthcare industry. It is a type of A-priori algorithm. It consists of following steps: collection of data, data customization, fine tuning, Splitting of data in key-value format, uploading data on Hadoop, determining the common item-sets, generation of candidates, determination of frequent item-sets and creation of rules for association. A global support parameter is used for determining strong rules of association. The binary partition of frequent item-sets is used to create rules from each frequent itemset. It is taking less time in transaction scanning than traditional A-priori. The Hadoop MapReduce scalable environment is used for implementation. The reduction in demand for resources as well as the reduction of overhead in communication is the main objective of this paper. This overhead is reduced in proposed approach by removing unusual data at an early stage. It is also producing lesser number of rules than existing algorithm.

Mingqi Lv et al. [4] proposed a novel pattern mining framework. This framework can adapt high degree of uncertainty present in dataset. The dataset contains personal trajectory data of people. The trajectory abstraction and the frequent pattern mining are two important parts of this framework. The initial step is preprocessing of trajectory data. It includes removal of outliers from data, reconstruction of data and compression of data. The next step is applying group-and-partition approach. The point-based trips are compressed and transformed as line-based trips. The final step is extracting route patterns using Spatial Continuity-based Pattern Mining (SCPM) algorithm.

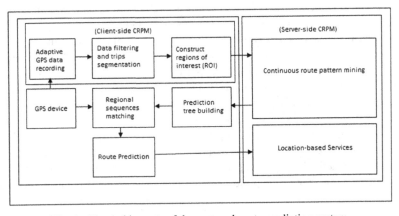

Fig. 1. The architecture of the personal route prediction system.

Ling Chen et al. [5] proposed a Client-Server Architecture which is shown in Fig. 1. It consists of three modules, namely, data preparation module, mining module and route prediction module. The first module is preparing data using five especially designed filters. The five designed filters are duplication filter, acceleration filter, angle filter, speed filter and total distance filter. The second module is mining route patterns using Continuous Route Pattern Mining (CRPM) algorithm. The third module is predicting route using the patterns mined from second module. The basic route mining algorithm and the heuristic route prediction algorithm are the decision tree-based prediction algorithms which are used to find route.

The different techniques related to pattern mining have been studied above and they are summarized in Table 1.

Table 1. Comparison study of pattern mining techniques.

Author	Year	Methodologies	Outcome	Pros	Cons
Frederic Stahl et al. [1]	2021	Fast Generalized Rule Induction (FGRI)	FGRI method constitutes a new kind of tool (data mining algorithm) in the Data Stream Mining toolbox	FGRI can handle both numeric and categorical data streams, it has been designed to be applied on tabular (symbolic) data streams	Induced ruleset is currently limited to 200 rules
Ming-Yen Lin et al. [2]	2021	Quantitative High-utility Item-sets mining (QHIM)	An efficient algorithm when the minimum value of threshold is low	The algorithm performs well for small data	Not suitable for large data
M. Sornal-akshmi et al. [3]	2021	Enhanced Parallel and Distributed A-priori (EPDA)	Algorithm being able to be used effectively to evaluate hidden patterns and produce related rules from datasets	Hadoop server is used that can handle many clients/nodes to implement master-slave cluster architecture	There is an overhead in processing
Mingqi Lv et al. [4]	2015	Spatial Continuity based Pattern Mining (SCPM)	Efficient and effective framework	Tolerating various kinds of disturbances in personal trajectory data handling high degree of uncertainty	Not Scalable

(*continued*)

Table 1. (*continued*)

Author	Year	Methodologies	Outcome	Pros	Cons
Ling Chen et al. [5]	2011	Continuous Route Pattern Mining (CRPM)	Extracting long route patterns	Extracting longer routes as compared to traditional substring methods	Performance can be improved using temporal attributes, transporta-tion means

A critical look at the available literature indicates the following issues which need to be addressed in future research.

- Algorithms are mining only short patterns.
- The algorithms extracting short patterns are not able to extract long patterns.
- The algorithms producing both short and long patterns but producing incomplete patterns.
- Some techniques are extracting only long patterns.

In this paper, a novel pattern mining algorithm has been proposed to address these issues. The extracted pattern or information obtained by applying this algorithm can be applied to various fields for proposing the solution of different problems.

3   Proposed Pattern Mining Algorithm

The Route Pattern Mining problem is defined as follows: Given a historical dataset in the form of trajectory and road network, designing a system that extracts frequent route item-sets in minimum time from huge trajectory dataset.

The objective of this work is to propose a novel algorithm from route prediction domain. The extracted route patterns after applying proposed approach can be viewed using Map Matching [15]. The meaningful insights extracted as a resultant of proposed approach can be further used to improve location-based services and to make intelligent transportation system services like estimating traffic congestion, route prediction, route planning and so on.

The Proposed algorithm for Frequent Route Pattern Mining (FRPM) is shown in Fig. 2. This algorithm extracts the frequent route patterns. The first step is to generate item-sets having cardinality one. The second step is to generate the item-sets having cardinality two. The third step is to generate item-sets having cardinality three. The generation of items of item-sets having cardinality three, requires the items of previous item-set i.e. items having cardinality two. In this paper, the items extracted in previous step is termed as *is_prev* which is further used to generate the items of next cardinality. The step 7 in Fig. 2 is generating all possible items from the dataset. It is using item-sets extracted in previous step to generate the item-sets of next iteration.

Definition 2: A *delete_first_data_point* (candidate item) is a function that gives a pattern obtained from the candidate item of itemset after deleting the prefix of cardinality one from it. The basic criterion for finding *delete_first_data_point* is the value of cardinality of candidate item should be at least two.

Let us suppose an itemset is given and this itemset contains k candidate item:

$$itemset = \{(P_{11}P_{12}P_{13} \ldots P_{1n})(P_{21}P_{22}P_{23} \ldots P_{2n}) \ldots (P_{k1}P_{k2}P_{k3} \ldots P_{kn})\}$$

Then, $delete\_first\_data\_point(P_{11}P_{12}P_{13} \ldots P_{1n}) = P_{12}P_{13} \ldots P_{1n}$.

where P_{im} denotes a GPS coordinate; Cardinality $|P_{im}| = n$; $n \geq 2$; m denotes the number of GPS points in a candidate item of an itemset, k denotes the number of candidate item in an itemset; GPS coordinate is in the form of (x_{ti}, y_{ti}, t_i); x_{ti}, y_{ti} and t_i denotes latitude, longitude and time respectively.

Algorithm 1: Frequent Route Pattern Mining (*dataset*)

1.   *itemsets*= ∅
2.   *is_all* = ∅
3.   generating candidates of cardinality one, *is_one*
4.   *itemsets*.insert(*is_one*)
5.   generating candidates of cardinality two, *is_two*
6.   *itemsets*.insert(*is_two*)
7.   generating candidates of all possible cardinality
    is_all= resultant itemsets(*is_two,dataset,support, is_all*)
8.   *itemsets*.insert(*is_all*)

Fig. 2. The proposed algorithm for route pattern mining

Definition 3: A *delete_last_data_point*(candidate item) is a function that gives a pattern obtained from the candidate item of itemset after deleting the suffix of cardinality one. The cardinality should be at least two.

Let, $itemset = \{(P_{11}P_{12}P_{13} \ldots P_{1n})(P_{21}P_{22}P_{23} \ldots P_{2n}) \ldots (P_{k1}P_{k2}P_{k3} \ldots P_{kn})\}$.
Then, $delete\_last\_data\_point(P_{11}P_{12}P_{13} \ldots P_{1n}) = P_{11}P_{12}P_{13} \ldots P_{1\,n-1}$.

where P_{im} denotes a GPS coordinate; Cardinality $|P_{im}| = n$; $n \geq 2$; m denotes the number of GPS points in a candidate item of an itemset, k denotes the number of candidate item in an itemset; GPS coordinate is in the form of (x_{ti}, y_{ti}, t_i); x_{ti}, y_{ti} and t_i denotes latitude, longitude and time respectively.

The iterative algorithm generating item-sets of all possible length is shown in Fig. 3. This algorithm is extracting patterns for short routes as well as for long routes. The algorithm of calculating support is shown in Fig. 4. This algorithm ensures credible and reliable route patterns are generated. A minimum threshold value of support is chosen to extract frequent patterns from the item-sets. The increase in value of minimum threshold will result in more frequent and lesser number of route patterns.

Algorithm 2: resultant itemsets (*is_prev, dataset, min_support, is_all*)

---

 1.   temp_seq= Φ
 2.   candidate_item= Φ
 3.   candidate_support= 0
 4.   *is_all*.insert(*is_prev*)
 5.   **for** item1= 0 to (*is_prev*.length-1) **do**
 6.   temp_item1= *is_prev*[item1]
 7.   **for** item2= 0 to (*is_prev*.length-1) **do**
 8.   temp_item2= *is_prev*[item2]
 9.   **if** (delete_first_data_point[temp_item1]==delete_last_data_point[temp_item2]) **then**
10.   candidate_item= temp_item1+ *is_prev*[item2][*is_prev*[item2].length-1]
11.   **end if**
12.   candidate_support= support(candidate_item)
13.   **if** (candiate_support≥ *min _support*) **then**
14.   temp_seq.insert(candidate)
15.   **end if**
16.   **if** (temp_seq is not empty) **then**
17.   resultant itemsets (temp_seq, *dataset, min_support, is_all*)
18.   **end if**
19.   **return** *is_all*

Fig. 3. The algorithm for generating candidate items of item-sets

---

Algorithm 3: support(*candidate, dataset*)

---

 1.   *support*= 0
 2.   item= Φ
 3.   **for** row= 0 to (*candidate*.length– 1) **do**
 4.   temp_sequence= Φ
 5.   **for** column= 0 to (*dataset*[row].length– *candidate*.length) **do**
 6.   **for** k=0 to (*candidate*.length–1) **do**
 7.   item= *dataset*[row][column+k]
 8.   temp_sequence.insert(item)
 9.   **if** (*candidate*==temp_sequence) **then**
10.   support= support+ 1
11.   **end if**
12.   temp_sequence.clear( )
13.   **return** *support*

Fig. 4. The algorithm for finding valid route patterns

4   Experimental Results

The evaluation of the approach is done using OpenStreetMap (OSM) dataset and Geo-Life dataset. OSM contains both spatial and non-spatial dataset. The non-spatial dataset includes length, width, name, speed, turn restriction and so on. The spatial data includes Centre line of roads, state boundaries, international boundaries and so on. The OSM data is shown in Fig. 5. The OSM data is licensed under open content licensed. This means it can be used free of cost. It is available in various format like.jpg, .jpeg, .png and so on.

GeoLife dataset contains time-stamped geographical information of around 182 users. It contains 24,876,978 GPS points; 17,621 trajectories; 1.2 million kilometers distance of trajectories; the total duration is approx. 48 thousand hours. These data are collected at a frequency of 1~5 s or 5~10 m per point.

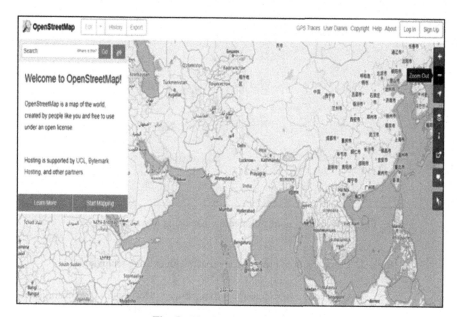

Fig. 5. The OpenStreetMap data

It is an open source dataset. An instance of raw GPS points is shown in Fig. 6. It contains only outdoor data. GeoLife dataset is collected during 2007 to 2012 from 182 users while they are performing their routine as well as non-routine tasks like site seeing, cycling, shopping, home to office, office to home and so on.

In Table 2, proposed approach FRPM is compared with existing methods based on certain parameters. It is a non-probabilistic approach. This can extract route patterns from trajectory of various length. The insights obtained from this can be used as a basis for further route prediction. Route prediction is the primary requirement of location-based services. By using the insights, various location-based services can be improved.

In Fig. 7, the experimental results are shown for five users. Similarly, the experiment is conducted for 182 users and the results were recorded and analyzed. The support parameter is used in this approach to decide the frequent patterns. By changing the value of support from one to five, results were recorded. From the result, it was concluded that the route patterns that were obtained as a result is frequent and can be further used in this domain of route prediction. The summary of the results obtained by varying support value is shown in Fig. 8. The Fig. 8 shows how support value decides frequent patterns. The analysis of experiment is that the increment in the value of support gives more frequent pattern and the number of frequent patterns reduces.

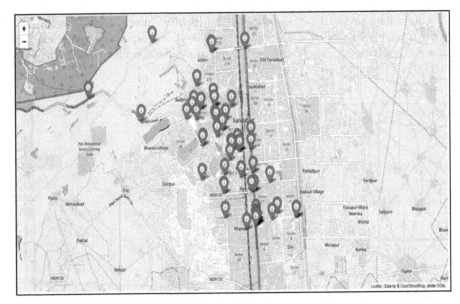

Fig. 6. Raw GPS points plotted on map using Map Matching

Table 2. Comparison of proposed approach with existing techniques.

Author	Approach	Extracting Short Patterns	Extracting Long Patterns	Horizontal Scalable	Probabilistic Model
Frederic Stahl et al. [1]	FGRI	Yes	No	No	No
Ming Yen Lin et al. [2]	QHIM	Yes	No	No	No
M. Sornalakshmi et al. [3]	EPDA	Yes	No	Yes	No
Mingqi Lv et al. [4]	SCPM	Yes	No	No	No
Ling Chen et al. [5]	CRPM	No	Yes	No	Yes
Proposed Approach	FRPM	Yes	Yes	No	No

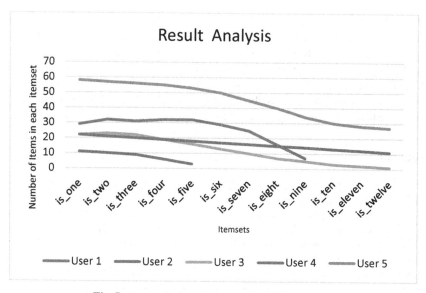

Fig. 7. Items obtained in itemset for different users

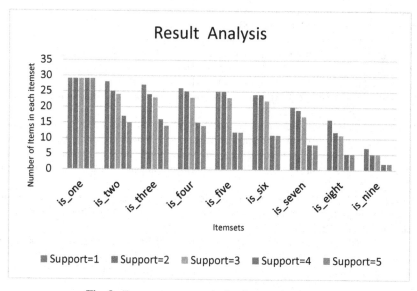

Fig. 8. Frequent patterns obtained on setting support

5 Conclusion and Future Scope

Pattern Mining is the process of extracting meaningful, useful patterns. It is having numerous applications in various domains. Different approaches proposed by different researchers had been applied in different domain. Each technique is having certain assumptions but extracting meaningful insights from them is a major concern.

The focus of this experiment is to propose a non-probabilistic pattern mining technique that can be applied in the domain of route prediction. The proposed pattern mining approach was applied on trajectory dataset to get meaning insights from people movement. A novel approach is proposed that can extract frequent route patterns. The main advantage of this approach is that it can extract patterns from routes of various length whether short or long. The insights obtained from this approach can be further applied on other applications of location-based services and further these services can be improved.

References

1. Stahl, F., Le, T., Badii, A., Gaber, M.M.: A frequent pattern conjunction Heuristic for rule generation in data streams. Inf. Switz. (2021)
2. Lin, M.-Y., Tu, T.-F., Hsueh, S.-C.: Mining high-utility itemsets of generalized quantity with pattern-growth structures. In: Peng, S.-L., Favorskaya, M.N., Chao, H.-C. (eds.) Sensor Networks and Signal Processing. SIST, vol. 176, pp. 447–464. Springer, Singapore (2021). https://doi.org/10.1007/978-981-15-4917-5_33
3. Sornalakshmi, M., et al.: An efficient apriori algorithm for frequent pattern mining using mapreduce in healthcare data. Bull. Electr. Eng. Inf. **10**, 390–403 (2021)
4. Lv, M., Li, Y., Yuan, Z., Wang, Q.: Route pattern mining from personal trajectory data. Proc. J. Inf. Sci. Eng. **31**, 147–164 (2015)
5. Chen, L., Lv, M., Ye, Q., Chen, G., Woodward, J.: A personal route prediction system based on trajectory data mining. Inf. Sci. **181**, 1264–1284 (2011)
6. Cao, H., Mamoulis, N., Cheung, W.: Mining frequent spatio-temporal sequential patterns. In: Proceedings of IEEE International Conference on Data Mining, pp. 82–89 (2005)
7. Giannotti, F., Nanni, M., Pedreschi, D.: Efficient mining of temporally annotated sequences. In: Proceedings of SIAM Conference on Data Mining, pp. 346–357 (2006)
8. Gidófalvi, G., Pedersen, T.B.: Mining long, sharable patterns in trajectories of moving objects. GeoInformatica **13**, 27–55 (2009)
9. Waghere, S.S., RajaRajeswari, P., Ganesan, V.: Retrieval of frequent itemset using improved mining algorithm in hadoop. In: Gupta, D., Khanna, A., Bhattacharyya, S., Hassanien, A.E., Anand, S., Jaiswal, A. (eds.) International Conference on Innovative Computing and Communications. AISC, vol. 1166, pp. 787–798. Springer, Singapore (2021). https://doi.org/10.1007/978-981-15-5148-2_68
10. Tiwari, V.S., Arya, A., Chaturvedi, S.: Route prediction using trip observations and map matching. In: Proceedings of Advance Computing Conference (IACC), 2013 IEEE 3rd International, pp. 583–587 (2013)
11. Minett, C., Salomons, M., Daamen, W., Arem, B., Kuijpers, S.: Eco-routing: comparing the fuel consumption of different routes between an origin and destination using field test speed profiles and synthetic speed profiles. In: 2011 IEEE Forum on Integrated and Sustainable Transportation Systems, FISTS (2011)
12. Azizi, M., Saniee Abadeh, M.: Protein structure prediction by means of sequential pattern mining. Proc. Int. J. Artif. Intell. Appl. **6**, 31–42 (2015)
13. Sohrabi, M., Azgomi, H.: Finding similar documents using frequent pattern mining methods. In: International Journal of Uncertainty Fuzziness and Knowledge-Based Systems (2019)
14. Kamal, M., Tahir, A., Kamal, M.: A survey for the ranking of trajectory prediction algorithms on ubiquitous wireless sensors. Sensors (2020)
15. Chao, P., Xu, Y., Hua, W., Zhou, X.: A survey on map-matching algorithms (2019)
16. Zheng, Y., Zhang, L., Xie, X., Ma, W.Y.: Mining interesting locations and travel sequences from GPS trajectories. In: Proceedings of WWW, pp. 791–800 (2009)

17. Zheng, Y., Xie, X., Ma, W.Y.: GeoLife: a collaborative social networking service among user, location and trajectory. IEEE Data Eng. Bull. **33**, 32–39 (2010)
18. Taguchi, S., Koide, S., Yoshimura, T.: Online map matching with route prediction. Proc. IEEE Trans. Intell. Trans. Syst. **20**, 338–347 (2018)

Student Clickstreams Activity Based Performance of Online Course

Anshu Singh[1]([⊠]) and Anuj Sachan[2]([⊠])

[1] University of Lucknow, Lucknow, India
[2] IIT-Roorkee, Roorkee, India

Abstract. MOOC is an online course targeted at massive-scale interactive involvement and web-based free access. The student prefers online video lecture. Analyzing the impact of student's interaction with different learning modes will improve the quality of an online course. The learning model used in this work is an online video course. In this work the authors have proposed features related to the number of time student watches a video, how much time spent on the course and the authors have taken help from secondary features like intermediate quiz performance and attendance in-class lectures. Using these features, the authors want to predict the performance of students in an online course. In this work, the author analyses students' video watching behaviour with the help of their on click event during the whole online course.

Keywords: MOOC · Moodle · Support vector machine · K-nearest neighbors

1 Introduction

The accessibility of the distance learning movement has gained much impetus over the last few years. Worldwide MOOCs- massive open online courses provide everyone accesses to education, people with a net connection can enroll in the various course offered by the different course provider (like Coursera).

Despite their convenience, the lower MOOC completion and commitment rates were criticized in comparison with traditional courses. Reasons behind the profound risk of dropout can be explained by the free nature of MOOCs student ability to self-regulate their learning or personal reasons. Besides, MOOCs face the universal educational challenge of keeping students engaged and motivated. The motivation behind the works is based on how students interact with an online course, their performance in that course, and why the dropout rate is high as the course proceeds. It gives the instructor an indication of which part of the course content is confusion and needs more explanation for that part.

2 Related Work

Various works have been done in the prediction of student performance in online courses. Some research is based on historical data, some of them use the clickstreams to get

© Springer Nature Switzerland AG 2021
A. Solanki et al. (Eds.): AIS2C2 2021, CCIS 1434, pp. 242–253, 2021.
https://doi.org/10.1007/978-3-030-82322-1_18

the behaviour of the student toward the video, some use the time-related property like viewing lectures at some point or plan to have long breaks before a new lecture begins for prediction of student performance. In [2] and [5] authors focus on the student video watching behaviour (e.g. recurring, play, fraction of time spent, skip, paused). [8] also studies video watching click-streams, except video-watching logs they used data from social networking (e.g., play and pause logs on the video, as well as post and comment threads on the forums).

In [3], the author studied time management behavior, student performance, and online learning association. The author has used features related to the viewing of lectures in some time or intends to delay before starting a new lecture for student prediction that is useful in identifying study habits and the prediction of performance.

In [6] a combination of student Week 1 work and social interaction is used by the author to predict their final performance in the MOOC course. The probability of students receiving the MOOC certificates and the certificate type (e.g. distinctiveness and normal) obtained can be predicted, too, with high precision. In [9], the author develops a system for predicting students' course grades for the next registration term using historical degree information.

3 Objectives

The objective of research is to observe the impact of the student interaction behaviour on the video lectures to their performance.

Objective 1: Identifying how student interact with video course, means how much time they spent on watching video lectures, number of times they watch video lectures (primary features) and intermediate quizzes performance, attendance of student are secondary features.

Objective 2: Prediction of student performance according to the features extracted in Objective 1.

4 Methodology

4.1 Data Pre-processing and Cleaning

After receiving data from the online learning platform, these data were processed. Remove all pairs of user-course that have no grade record. Then the authors dropped the pairs with a grade record but with an unfinished trajectory of the user course. Also, Those records have been removed from our datasets with at least one "null" parameter entry. Remove outlier using box-plot so that they cannot affect feature-offset.

4.2 Feature Extraction

Obtained different types of data from video watching logs and course database: video watching frequency means the number of time particular video has been watched, each video how many times watched by each student, total time spent on a course by each student, intermediate quiz performance result, and attendance of each student. Using these above data features are formulated.

First Feature. In this first feature, Score has been formulated for each student using video watching frequency (how the number of time students watches a video) and the Weight (importance given by the students to a particular video) of videos.

Determining Importance of a Lecture. The importance of video lecture has been calculated based on each student's number of times each video lecture watched. For calculation of video importance following three parameters are taken into account. Timestamp Logs are used to construct the portion of the video that the student has watched. **First parameter** is the initial state of the video after being click by the student should be an un-started event noted by a plugin into the database student activity logs through which videos areas been uploaded in server. **Second parameter** is the student has to watch at least 7 min of a particular video, so that it should be considered that the student has watched a particular video. **Third parameter** is at the end of the video state of the video should be ended.

Weight Assigning to Videos. The authors have divided videos based on their view count in 10 different class. Weight is assigned to these ten different classes based on the number of views. Given the maximum number of views can be 500, the authors have clubbed the number of views in an increasing interval of 50, which are assigned values from 1 to 10, respectively. This value represents the Weight of the video. (importance of video).

Score Feature Calculation. Calculation based on Weight and view count. Suppose the number of video lecture is K. Student start with i = 1 to 150 (number of students) and j = 1 to K for the number of video lecture.

$$\sum_{j=1}^{K} v_j^i w_j \text{ for every i } = 1 \text{ to } 150; \tag{1}$$

where $v^i{}_j$ represent number of time j^{th} video view/watched by i^{th} student. w_j represent weight corresponding to particular lecture. (Importance given by student through viewing). Calculating $v^i w_j$ for each student. For example student id-1 calculate

$$v_1^1 w_1 + v_2^1 w_2 + v_3^1 w_3 + v_4^1 w_4 + \ldots + v_{43}^1 w_{43}. \tag{2}$$

Second Feature. In this section, the second feature (watching time) is formulated using the students' timestamp activity logs on the course. The second feature is related to time spend on a course. For example, for a video, calculate the total time spent by subtracting the adjacent timestamp and then sum the difference, which will give us the total time spent by a student on a video. If the authors observe a timestamp related click-play event of t sec followed by another timestamp event t + 10 s, the authors can calculate the user who viewed the video for 10 s. Another click event occurs with timestamp p + 15 s, then the adjacent difference of timestamp is 10 s. After that, the total time spent on video is 5 + 10 s. A similar procedure is done for each video concerning each student to calculate the watching time feature for each student.

Third and Fourth Feature Extraction. In this section, the third and fourth features are extracted using the quiz performance on uploaded quiz related to the course and attendance in class for each student.

4.3 Dataset Split and Evaluation

In this section, the dataset of 150 students is divided into two-part. 70% of the available data used for classifier training and to test their performance, the authors, have remaining 30% of data. In this, the authors have four groups: Poor, Average, Good, and Excellent. All the 150 students belong to one of the group. The authors used two algorithms to train two classifier for the validation of the new feature proposed. Classification Algorithm used are [11] support vector machine, [12] K-nearest neighbors.

5 Experiment and Analysis

This section discusses the experimental setup, the dataset used and the dataset dimension. Various ways to look at the data using univariate and multivariate plots to better understand the data. Experiments on the dataset using k-nearest neighbours (KNN) and Support Vector Machine (SVM) and discuss their findings. The classification methods attempt to predict the categorical outputs in the data sets obtained by experimental methods by mathematical methods. To predict which of the previously unknown values belong to the classes defined by the experiments, the discovery of the mathematical relation between the variables with input values and the class vector with output values is used [21]. K-nearest neighbour (KNN) and support vector machine (SVM) algorithms are used commonly in classification. The KNN algorithm is one of the most popular classification algorithms. The most basic [22], the SVM algorithm has been implemented in different studies. It has been chosen to be efficient classification methods with the high-performance generalization as a result of these studies [23].

5.1 Setup

In the Experimental setup, a moodle 2.9+ version was installed on Ubuntu Server 14.04 to create an online computer networks course for IIIrd semester students. The authors added the plugin named 'vidtrack' on Moodle through which videos were uploaded on the Moodle server. When a user clicks on the video, the following information are stored in the database. Table 2 shows the description of the course created on moodle (Table 1).

Table 1. Record table

id	Unique
user	user id
course	course id
video	vidtrack id
state	Captured Event
time	Time at which event

Table 2. Course description

Course	Learners	Video lectures	# Events	Avg. video length (min)
CN	150	43	247564	15 min

Student clickstreams log contain events like play, pause etc. information captured when student view video lectures are shown in Table 3.

Table 3. Event description

Type of event	Description	# Event
Playing	Users click the play button it will play automatically and a play event will be recorded	112676
Paused	Users click the pause button, a pause event will be recorded	39111
Buffering	Due to buffering video stalled	88547
Unstarted	video has not been watched	7230

5.2 Summary of the Dataset

In this step the authors will take a few different ways to look at the data: Dataset dimension, Summary statistics of all the attributes and Visualization of data. In **dataset dimension**, the authors will find out how many rows (instances) and columns (attributes) in our dataset. Dimension of dataset is (150×4).

Statistical summary of dataset is described in Table 4. A summary of each attribute can be viewed. The count, mean, standard deviation, min and max values as well as some percentiles are included.

Table 4. Statistical summary

	Score	Quiz	Attendance	Watching time (min)
count	150.000000	150.000000	150.000000	150.000000
mean	644.940000	7.626667	10.660000	486.413333
std	328.352512	2.395596	3.243641	222.862951
min	0.000000	0.000000	0.000000	0.000000
25%	389.250000	6.000000	9.000000	333.250000
50%	725.500000	9.000000	12.000000	568.000000
75%	855.500000	9.000000	13.000000	646.500000
max	1473.000000	10.000000	15.000000	834.000000

Class distribution section, the authors will see how many instances belong to each class. Table 5 shows the class distribution of dataset. For each class, it is showing the absolute count. This class distribution table shows us that data is imbalanced data. Imbalanced data means each class label has a different absolute count. Both the Table 4 and 5 obtain by evaluating student data on Python platform.

Visualization of Data. The authors have a fundamental idea about the data now. With some visualizations, I need to extend that. The authors look at two kinds of plots:

Univariate plots, to better understand each attribute. In Fig. 1 the authors begin with some univariate plots, that is, each variable's plots. Since the input variables are numeric, each of the whisker and box plots can be created. A plot of boxes and whiskers is a way to summarize a data set of intervals. The box edges are drawn at the data's 25th and 75th percentiles, and the 50th percentile is marked by a line in the middle of the box. Whiskers are the two outboard lines extending to the highest and lowest observations.

Table 5. Class distribution

Class	Poor	Average	Good	Excellent
	31	31	70	18

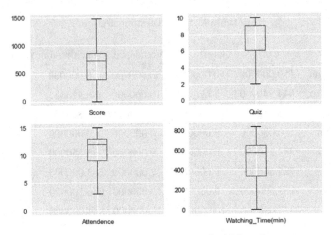

Fig. 1. Univariate Plots- box and whisker plots

The authors can create a histogram of each variable input to get a sense of the distribution. Histograms are a type of bar chart showing the numbers or relative frequencies of values falling at different intervals or ranges of classes in Fig. 2.

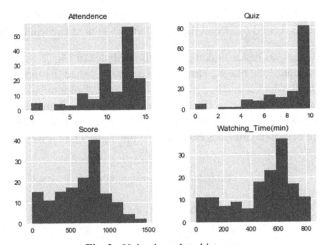

Fig. 2. Univariate plots-histogram

Multivariate plots, in order to better understand the attributes relation- ships. Figure 3, first look at all pairs of scatter plots of attributes. It can be useful in spotting relationships of structured input variables.

5.3 Evaluation Metrics

The dataset is an unbalanced dataset because the authors have a different count for the different group (category), so for reporting the classifier's performance, the authors have used three metrics: precision, recall, and f1-score. In addition, for each class, the authors have shown each of these metrics. Our metrics defined below:

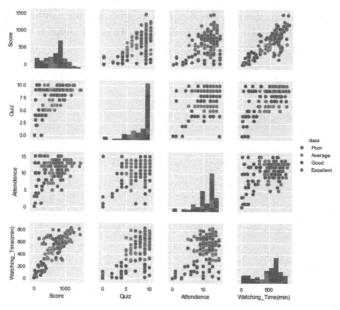

Fig. 3. Multivariate plots

– Precision(P) – Precision basically tells us how many were actually positive out of the results our model classified as positive.

$$Precision(P) \ = \ \frac{(TP)}{(TP \ + \ FP)} \qquad (3)$$

Recall(R)–(sensitivity) - Recall tells us how many true positives our model has recalled or found (points labeled as positive).

$$Recall(R) \ = \ \frac{(TP)}{(TP \ + \ FN)} \qquad (4)$$

f1-score – which is defined as the harmonic mean of precision and recall. It tells us about the balance between precision and recall. Some models may be high- recall model and some models may be high-precision models.F1-score is a metric takes into account both precision and recall as the authors can't constantly evaluate both and take the higher one for our model.

$$f1 - score = 2 * \frac{(P * R)}{(P + R)} \qquad (5)$$

5.4 Results and Discussion

Table 6 list the classification performance of SVM classifier using our features. For each feature, the authors have shown all the metrics. For SVM precision and recall for all feature together is 0.82 and 0.82 respectively. f1-score is calculated by the formula given above is 0.80. Maximum accuracy achieved is 82.22% Table 7 list the classification performance of K-nearest neighbours classifier using our features. For each feature, the authors have shown all the metrics. For K-nearest neighbours, precision and recall for all feature together are 0.87 and 0.87 respectively. f1-score is calculated is 0.86. Maximum accuracy achieved in K- nearest neighbours is 86.67%

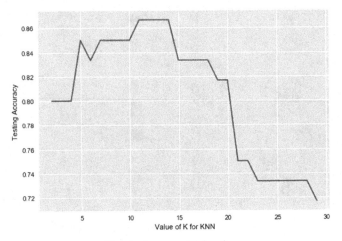

Fig. 4. Accuracy vs k-value

Comparing the model concerning f1-score because data is imbalanced, K- nearest neighbour performs better than the support vector machine. K-nearest neighbours give f1-score 0.86, and support vector machine gives f1-score 0.80. **K-Nearest Neighbors Result Visualization** In Fig. 4, the authors plot Testing accuracy over the value of K-nearest neighbours. Range of k value from 1 to 30. It shows that initially, when the k value is less, the testing accuracy is also low while increment k value, testing accuracy also increases. Maximum testing accuracy achieved when the k value reaches 11. It maintains there accuracy till k value 14 after that, accuracy decreases, then the accuracy achieved at k equal to 11. Maximum Testing accuracy achieved is 87%.

Table 6. SVM classification report

Class	Precision	Recall	f1-score	Support
Average	0.80	0.40	0.53	10
Excellent	0.80	0.80	0.80	5
Good	0.77	0.94	0.85	18
Poor	0.92	1.00	0.96	12
Avg/total	0.82	0.82	0.80	45

Table 7. KNN classification report

Class	Precision	Recall	f1-score	Support
Average	0.86	0.60	0.71	10
Excellent	0.80	0.80	0.80	5
Good	0.81	0.94	0.87	18
Poor	1.00	1.00	1.00	12
Avg/total	0.87	0.87	0.86	45

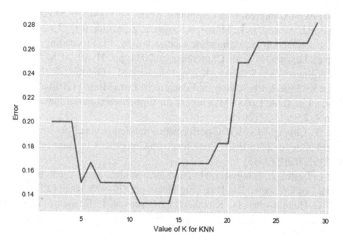

Fig. 5. Error vs k-value

In Fig. 5, the authors' plot Error over the value of K-nearest neighbour. It shows at what point of k error is minimum. When the k value is 11, it showed a minimum error.

6 Conclusion

In this work, performance prediction has been based on four features. The first two primary features include the number of times a particular video has been watched and the total time spent on the course by students. The other two are secondary features which include attendance and performance in intermediate quizzes of students. With dataset from our online Moodle learning Platform and plugin, two goals have been accomplished: (1) Novel features for describing learning behaviour, which is useful in identifying performance prediction has been presented (2) Final grades of students has been predicted exclusively based on features extracted. Performance prediction has been done using two classifiers, i.e., SVM (Support Vector Machine) and KNN (K-Nearest Neighbours). For the four extracted features, KNN predicts with high accuracy of 86.67% as compared to the accuracy of 82.22% for SVM.

References

1. Shridharan, M., Willingham, A., Spencer, J., Yang, T.Y., Brinton, C.: Predictive learning analytics for video-watching behavior in MOOCs. In: 2018 52nd Annual Conference on Information Sciences and Systems (CISS), pp. 1–6. IEEE, March 2018
2. Brinton, C.G., Buccapatnam, S., Chiang, M., Poor, H.V.: Mining MOOC clickstreams: video-watching behavior vs. in-video quiz performance. IEEE Trans. Signal Process. **64**(14), 3677–3692 (2016)
3. Hong, B., Wei, Z., Yang, Y.: Online education performance prediction via time-related features. In: 2017 IEEE/ACIS 16th International Conference on Computer and Information Science (ICIS), vol. 64, no. 14, pp. 95–100. IEEE (2017)
4. Kovacs, G.: Effects of in-video quizzes on MOOC lecture viewing. In: Proceedings of the Third (2016) ACM Conference on Learning@ Scale, pp.31–40. ACM (2016)
5. Kim, J., Guo, P.J., Seaton, D.T., Mitros, P., Gajos, K.Z., Miller, R.C.: Understanding in-video dropouts and interaction peaks inonline lecture videos. In: Proceedings of the First ACM Conference on Learning@ Scale Conference on 2014, pp. 31–40. ACM (2014)
6. Jiang, S., Williams, A., Schenke, K., Warschauer, M., O'dowd, D.: Predicting MOOC performance with week 1 behavior. In: Educational Data Mining (2014)
7. Chang, C.-C., Lin, C.-J.: LIBSVM: a library for support vector machines. ACM Trans. Intell. Syst. Technol. **2**(3), 1–27 (2011)
8. Brinton, C. G., Chiang, M.: MOOC performance prediction via click- stream data and social learning networks. In: 2015 IEEE Conference on Computer Communications (INFOCOM), pp. 2299–2307. IEEE (2015)
9. Sweeney, M., Lester, J., Rangwala, H.: Next-term student grade prediction. In: 2015 IEEE International Conference on Big Data (Big Data), pp.970–975. IEEE (2015)
10. Anderson, A., Huttenlocher, D., Kleinberg, J., Leskovec, J.: Engaging with massive online courses. In: Proceedings of the 23rd International Conference on World Wide Web on 2014, pp. 687–698. ACM (2014)
11. Suykens, J.A.K., Vandewalle, J.: Least squares support vector machine classifiers. Neural Process. Lett. **9**(3), 293–300 (1999). https://doi.org/10.1023/A:1018628609742
12. Guo, G., Wang, H., Bell, D., Bi, Y., Greer, K.: KNN Model-Based Approach in Classification. In: Meersman, R., Tari, Z., Schmidt, D.C. (eds.) OTM 2003. LNCS, vol. 2888, pp. 986–996. Springer, Heidelberg (2003). https://doi.org/10.1007/978-3-540-39964-3_62
13. Graf, H., Cosatto, E., Bottou, L., Dourdanovic, I., Vapnik, V.: Parallel support vector machines: the cascade SVM. Advances in Neural Information Processing Systems on 2004, pp.521–528 (2004)
14. Auria, L., Moro, R.A.: Support vector machines (SVM) as a technique for solvency analysis. DIW Berlin Discussion Paper on 2008 (2008)

15. Kloft, M., Stiehler, F., Zheng, Z., Pinkwart, N.: Predicting MOOC dropout over weeks using machine learning methods. In: Proceedings of the EMNLP 2014 Workshop on Analysis of Large Scale Social Interaction in MOOCs on 2014, pp. 60–65 (2014)
16. Chen, Z.-C.: Automatic self-feedback for the studying effect of MOOC based on support vector machine. In: Sun, X., Pan, Z., Bertino, E. (eds.) Artificial Intelligence and Security: 5th International Conference, ICAIS 2019, New York, NY, USA, July 26-28, 2019, Proceedings, Part II, pp. 309–320. Springer International Publishing, Cham (2019). https://doi.org/10.1007/978-3-030-24265-7_27
17. Borrella, I., Caballero-Caballero, S., Ponce-Cueto, E.: Predict and intervene: addressing the dropout problem in a MOOC-based program. In: Proceedings of the Sixth (2019) ACM Conference on Learning@ Scale on 2019, pp.1–9. ACM (2019)
18. Pérez-Sanagustín, M., Sharma, K., Pérez-Álvarez, R., Maldonado-Mahauad, J., Broisin, J.: Analyzing learners' behavior beyond the MOOC: an exploratory study. In: Scheffel, M., Broisin, J., Pammer-Schindler, V., Ioannou, A., Schneider, J. (eds.) EC-TEL 2019. LNCS, vol. 11722, pp. 40–54. Springer, Cham (2019). https://doi.org/10.1007/978-3-030-29736-7_4
19. Pham, P., Wang, J.: AttentiveLearner: improving mobile MOOC learning via implicit heart rate tracking. In: Conati, C., Heffernan, N., Antonija Mitrovic, M., Verdejo, F. (eds.) Artificial Intelligence in Education, pp. 367–376. Springer International Publishing, Cham (2015). https://doi.org/10.1007/978-3-319-19773-9_37
20. Zhang, H., Yang, H., Huang, T., Zhan, G.: DBNCF: Personalized courses recommendation system based on DBN in MOOC environment. In: 2017 International Symposium on Educational Technology (ISET), pp.106–108. IEEE (2017)
21. Frank, I.E., Friedman, J.H.: Classification: oldtimers and newcomers. J. Chemometr. 3(3), 463–475 (1989). https://doi.org/10.1002/cem.1180030304
22. Altay, O., Ulas, M.: The use of kernel-based extreme learning machine and well-known classification algorithms for fall detection. In: Bhatia, S.K., Tiwari, S., Mishra, K.K., Trivedi, M.C. (eds.) Advances in Computer Communication and Computational Sciences: Proceedings of IC4S 2017, Volume 2, pp. 147–155. Springer Singapore, Singapore (2019). https://doi.org/10.1007/978-981-13-0344-9_12
23. Wang, H., Zheng, B., Yoon, S.W., Ko, H.S.: A support vector machine-based ensemble algorithm for breast cancer diagnosis. Eur. J. Oper. Res. 267(2), 687–699 (2018)

Feature Selection with Random Forests Predicting Metagenome-Based Disease

Huong Hoang Luong[1] , Thanh Huyen Nguyen Thi[1] , An Duc Le[1],
and Hai Thanh Nguyen[2](✉)

[1] FPT University, Can Tho, Can Tho 900000, Vietnam
[2] College of Information and Communication Technology,
Can Tho University, Can Tho, Vietnam
nthai.cit@ctu.edu.vn

Abstract. The early detection of at-risk human diseases is a significant challenge for health care professors. Data Volumes related to the disease are usually relatively large, but the capability of human computation is limited. This study aims to provide a solution for disease prediction by selecting characteristics of the original set of features on metagenomic data. We propose a novel approach to enhance the prognosis of inflammatory bowel disease (IBD) and colorectal cancer (CRC), which are to use Random Forests and the Support Vector Machine (SVM). The results with the selected features using the proposed method are pretty promising on datasets of Colorectal Cancer and Inflammatory Bowel Disease (IBD) compared to the original set of features using state-of-the-art techniques.

Keywords: Feature selection · Random forest · Support vector machine · Metagenomic data · Colorectal cancer

1 Introduction

Metagenomics is a term that was formed first in 1998 [1] and later popularized. It is a method of applying genomic sequencing or assay functional characteristics to analyze not dependent on culture microbial populations' complex and diverse. Metagenomics is a method of analyzing multi-genomic DNA [2] (DNA metagenome) of all microorganisms obtained directly from natural environment samples [3]. In essence, Metagenomics is the term used to describe a field of scientific research and techniques that allows analyzing all microorganisms living in any natural environment [4]. The benefit it brings is acquiring and analyzing all the genomes richness and diverse microorganisms, especially 99% of microorganisms without culture [5]. Metagenomics becomes the tool to help the researchers the most effective exploitation of new genetic resources. Raw data are processed by biology methods in order to acquire the composition of species, functional features, and richness [6,7]. A sample measured is described by the microbial taxonomy, which may be the relative abundance of microbial units including domain, kingdom, phylum, class, order, family, genus, or species. It is used to

© Springer Nature Switzerland AG 2021
A. Solanki et al. (Eds.): AIS2C2 2021, CCIS 1434, pp. 254–266, 2021.
https://doi.org/10.1007/978-3-030-82322-1_19

determine the relative abundance of bacteria and to associate it with host diseases that would allow us to improve its early diagnosis. This data may also provide insight into the mechanism of the disease. Nevertheless, the association of individual bacteria with a particular disease type is inconsistent due to various problems such as the complexity of the disease, and the limited amount of observed data. Besides, biological and metrological data are quite complex multidimensional data and challenge for facing humans. In recent years, significant improvements have been made in measurement. The calculation must be done correctly. That is the reason why machine learning has come into common use in solving metagenomics problems. These are the core issues mentioned in [8]. Operational taxonomy units (OTU) [17] such as subgrouping, classifying, classifying and assigning, comparative measurement and prediction of genes. The need to use optimization techniques to exploit the omics dataset is a challenge because of the huge amount of features and the relative number of observations. Machine Learning (ML) [9] techniques make it easier to analyze the data, then apply metagenomics. In our study, metagenomics and ML were used to find and early detect the carriers of inflammatory bowel disease (IBD) [10,11] and colorectal cancer (CRC) [12].

In this paper, we operate an approach using Random Forest to predict Colorectal cancer and Inflammatory bowel disease. The aim of this paper is to provide an efficient method base on reliable predictive results. Our paper is divided into four main sections. In Sect. [2], we mention points related to study work. In the Sect. [3], we detail the description of the method used in this paper. In the Sect. [4], we study in detail and show the experiment. In the final Sect. [5], we conclude the main points in this article.

2   Related Work

Many research studies have applied Machine Learning to evaluate and visualize metagenomic [13] datasets. Metagenomic data is a new source of data to improve diagnosis and prognosis for patients. DL has achieved a lot of success in measuring issues related to OTU [17]. This paper introduces one-dimensional 1D based binning method and scaling algorithms that improve predictive ability and performance for diseases based on metagenomic data using artificial neural networks. Assessment method on 7 datasets related to 6 diseases: Cirrhosis of the liver [18], colorectal [19], cancer, IBD [10], type 2 diabetes, obesity, HIV. The heart disease CMD leads to the chronic stages of heart and heart failure. This method diagnoses and prognoses for a long time. They apply this method of treatment to patients with similar diagnoses. Some patients have improved health, some have not. ML is generally DL providing tools and algorithms to propose new models and methodologies and at the same time place hypothesis testing and therapeutic products. This study also uses Multi-layer Perceptron (MLP) and traditional artificial neural networks. In the paper, they aim to divide the bin into different which includes width, frequency and proposed breaks. After that, multi-fold (K-fold) is used to classify diseases in the process of learning with MLP. The Binning classification method was implemented with 10 bins. The implemented MLPs are the Rectified Linear Unit (ReLU). In short, proposed Met2Bin with various approaches using binning and proportional analysis methods observed for metagenomic. The binning method outperforms six of the seven standard datasets, improving performance [14].

Moreover, the authors explore many methods relate to colorectal cancel for visualizing features of metagenomic datasets. In order to visualize data distribution and use for classification, they proposed three methods including Image Generation and Models for Evaluation, Visualizations Based on Dimensionality Reduction Algorithms, and Visualization Based on Data Density. Image Generation and Models for Evaluation use Species Bins and Quantile Transformation Bins to generate the image, after that, they use CNN to evaluate the considered visualization and models. Visualizations Based on Dimensionality Reduction Algorithms apply dimensionality reduction algorithms like Principal Component Analysis, t-SNE, Random Projection, Isomap, Spectral Embedding, Locally Linear Embedding, Multidimensional scaling and Non-negative matrix factorization. These algorithms are used to shaped high-dimensional data into 2D images for improving prediction results. Visualizations based on the Data Density represent the data depend on the data density with Fill-up [15]. Further, data visualization is still a challenge for many countries in the world. With measurement data, it is known by the size of the data which is tough for humans to understand. For some diseases that use measurement data to predict, DL is regularly less effective than Classical Machine Learning in predicting CRC. Authors represent a method using Manifold Learning with t-SNE [16] and Spectral embedding to display digital data into images and using DL to improve performance in predicting CRC.

3   Methodology

In this section, we explain how to select features. A method we use to select features using Random Forest. It is a decision tree algorithm, with numbers in the hundreds. Each decision tree is generated randomly from sample re-selection (bootstrap, random sampling) and using only a small set of random features from all variables in the data. In the final state, the RF model usually works very accurately and it is a well-known technique. Due to the effects of the method, we use RF for our study. Next, we sort them incrementally and select 100 features that have correlation which is the greatest in datasets for later comparison. We use 100 features including 2 reasons. First, 100 features were just enough for us to test out. Second, our main purpose is to demonstrate a reduction in the number of features while still delivering better results and performance. After that, using two approaches which are Random Forest and Support Vector Machine to predict results from selected features by performing a loop incrementally from 1 to 100 features for each time. Using RF and SVM methods are we compare their effectiveness and prove that Random Forest gives better results. The prediction is implemented by the specific number of features and primarily based on a dataset archived in the study of Fungal microbiota dysbiosis in CRC and IBD. The main aim of us is to minimize the number of features to predict disease better.

3.1   Feature Selection Using Random Forest

In our study, we select feature using Random Forest (RF) [20,21]. RF is also known as Random Decision Forest which consists of 4–12 hundred decision trees. Trees are constructed by a random extraction of the recorded information from the dataset and a random extraction of the features. Each tree is also a series of yes-no questions based on one or a combination of features. At each node is at each question, the three splits the dataset into two groups, each of them containing observations that are comparable among themselves and dissimilar to observations in the other groups. Hence, the important feature is purely selected from each of the groups.

The datasets as mention in Sect. [4.1], they consist of 2 datasets. Each is corresponding 4 and 6 sets of data. The number of rows in one set delegated the subjects. Those pursuing columns are the code of the bacteria and their value corresponding to the subjects. After implementing method RF to find out important features. Then, we sort them incrementally by value and get the first 100 features.

3.2   Predict Using Random Forest on RF

As we presented in Sect. [3.1], Random Forest is a set of Decision Trees. Each node of the tree is properties, and branches are the selected value of that property. Following the property values on the tree, a decision tree tells us the predicted value. Thus, this is a method of building a set of lots of decision trees and implement voting to make decisions about the target variable that needs to be predicted. After selecting 100 important features using RF, we get the features on the test data and predict the results by Decision Trees, the incremental features from 1–100. It calculates the number of votes in the entire forest for each result. The end result is based on the metric mention in Section [4.2].

3.3   Predict Using Support Vector Machine on RF

The Support Vector Machine (SVM) [22,23] is a categorical approach. It can use to handle many contiguous variables and classifications. A hyperplane is built by SVM in multidimensional space to separate different classes. SVM creates optimal hyperplane in a repetitive way to minimize errors. The aim of SVM is to define a marginal hyperplane which uses to divide into the best data set classes. We get 100 features which are selected in Sect. [3.1]. Using SVM to predict calculates results ACC with the incremental features from 1–100.

4   The Experiments

4.1   Dataset Description

The IBD dataset was published in [24] for a study about Fungal microbiota dysbiosis in IBD. In that study [24], the authors got bacterial and fungal from 38 healthy subjects (HS) and 235 patients were resolved using 16S and ITS2 sequencing. We use two datasets.

- The first data is CRC (Colorectal Cancer) dataset which has been used on a paper [25]. It includes 4 sets of data such as Feng, Vogtmam, Yu, Zeller.
  - Vogtmam - Cohort C1
  - Feng - Cohort C2
  - Yu - Cohort C3
  - Zeller - Cohort C4
- The second data is the IBD dataset which has been announced on 6 September 2017 [26] afforded the data consist of 6 sets of data as follows:
  - HS_UCr - Healthy (HS) and Ulcerative Colitis (UC) in remission.
  - HS_iCDr - Healthy (HS) and ileal Crohn's disease (iCD) in remission.
  - HS_UCf - Healthy (HS) and Ulcerative Colitis (UC) in flare.
  - HS_iCDf - Healthy (HS) ileal Crohn's disease (iCD) in flare.
  - HS_CDr - Healthy (HS) Crohn's diseasen(CD) in remission.
  - HS_CDf - Healthy (HS) Crohn's diseasen(CD) in flare.
  Information detail are showed in the Table [1].

Table 1. Datasets information detail

Information	CRC				IBD					
	Feng	Vogtmann	Yu	Zeller	UCr	iCDr	UCf	iCDf	CDr	CDf
Total features	1978	1976	1932	1976	238	258	251	248	258	260
Patients	46	48	73	88	44	59	41	44	77	60
Healthy subject	63	52	92	64	38	38	38	38	38	38
Total subjects	109	100	165	152	82	97	79	82	115	98

4.2   Data Division and Scoring Metrics

We have been known two informative datasets. Moreover, we divide those datasets into two groups: the first group for the train and the second group for tests. In the CRC dataset, the first group we choose Feng data to select features afterward for the train. In other group, four sets others for test. In IBD dataset implements UCr for train and the others for test.

Using score metrics to evaluate the results is Accuracy (ACC). It has value in a range from 0.0 to 1.0. The value of 1.0 indicates perfect prediction and a value of 0.0 means that the prediction false. To calculate the ACC that depends on the number of correct predictions $n_{correct}$ and total predictions n_{total}, we have an accuracy equation be:

$$ACC = \frac{n_{correct}}{n_{total}}$$

4.3 Prediction Result with ACC Metrics

In two prediction methods, we got 100 features for the train and test. It means we operated a loop it runs 1–100 times. Each time, the number of features was executed corresponding to the number of iterations. For example, the first time gets 1 feature for the train and test. In the figures showing the results of 100 features, but due to a large number of features, we showed the results represented in Tables 2 and 4 as 10 features.

The ACC results are executed by using two chart. There are the Boxplot chart which depicts distribution and deviation of values and the Line chart which depicts change of values. Two chart are divided into 2 parts, there are Y-axis is accuracy (ACC) and X-axis is number of features. In Table 2, it includes result of Figs. 1 and 2 using Random Forest for prediction. In Table 4 shows results of two figures such as Figs. 3 and 4 using Support Vector Machine for prediction.

Table 2. Results of ACC with predicting uses Random Forest (RF)

Datasets	Features										
	Name	1	2	3	4	5	6	7	8	9	10
CRC	Yu	**0.6424**	0.5394	0.5636	0.5636	0.5939	0.6000	0.5879	0.5697	0.6000	0.5879
	Zeller	0.5197	0.5263	0.5789	0.5921	0.6053	0.5987	0.6645	0.6776	0.6645	**0.6711**
	Vogtmann	**0.6400**	0.5100	0.5400	0.5300	0.5500	0.5400	0.5400	0.5500	0.5400	0.5200
	Average	0.6007	0.5252	0.5608	0.5619	0.5831	0.5796	0.5975	0.5991	**0.6015**	0.5930
IBD	iCDr	0.8454	0.8763	0.8866	0.8969	0.9175	0.9072	0.9072	0.9072	**0.9278**	0.9278
	UCf	0.8228	0.8861	0.9114	0.9114	**0.9367**	0.9367	0.9367	0.9367	0.9367	0.9367
	iCDf	0.8537	0.8902	0.9512	0.9512	0.9512	**0.9634**	0.9512	0.9512	0.9634	0.9512
	CDr	0.8174	0.8696	0.8783	0.8783	0.9043	0.8957	0.8957	**0.9130**	0.9043	0.9130
	CDf	0.8163	0.8776	**0.9592**	0.9490	0.9490	0.9592	0.9592	0.9490	0.9592	0.9490
	Average	0.8311	0.8800	0.9173	0.9174	0.9317	0.9324	0.9300	0.9314	**0.9383**	0.9355

From Table 2, it shows the results as predicting using Random Forest. ACC results are represented from 1 to 10 features. The result includes 2 datasets. On CRC without number of features, the highest ACC values in the dataset (Yu, Zeller, Vogtmann) are 0.6424, 0.6711, and 0.6400 respectively. The average is 0.6015. To IBD, the highest ACC values in the dataset (iCDr, UCf, iCDf, CDr, CDf) are 0.9278, 0.9367, 0.9634, 0.9130 and 0.9592 respectively. The average is 0.9383.

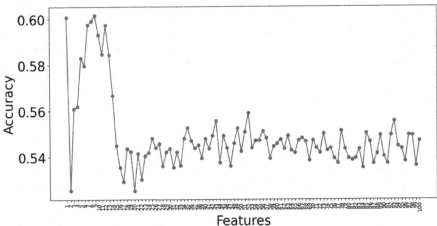

(a) Average ACC on Line Chart.

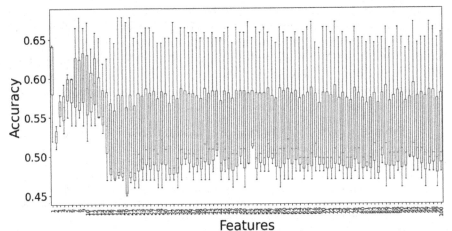

(b) The distribution and skewness of ACC on Boxplot Chart.

Fig. 1. Predict using Random Forest on CRC dataset (100 features) (Color figure online).

Figure 1, it depicts using Random Forest method to predict on CRC dataset. It is divided into 2 graphs. First graph [1a], The Line represents average ACC and maximum value is **0.6015**. There are 100 features. The alternative number is much from 1 to 35 features, remaining it seems only change a little. Second graph [1b] is called Boxplot chart which illustrates a set of values achieved (ACC) such as the minimum, the maximum, the sample median, and the first and third quartiles. The graph swiftly rises and falls from 1 to 22 features. The highest values is **0.6776**.

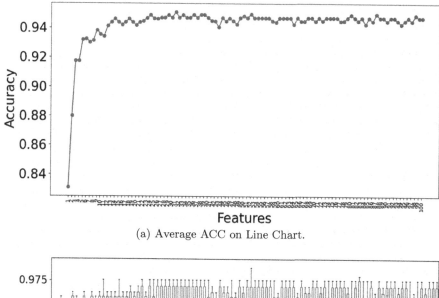

(a) Average ACC on Line Chart.

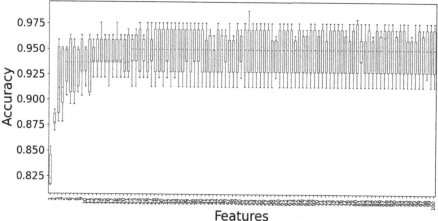

(b) The distribution and skewness of ACC on Boxplot Chart.

Fig. 2. Predict using Random Forest on IBD dataset (100 features) (Color figure online).

The method using to predict the same with Fig. 1 that is Fig. 2 on IBD dataset, The Line [2a] has values from 1 to 14 features is gradually increasing and the highest ACC is **0.9508**. Values afterward are maintained. The next graph is known the Boxplot [2b] which have a growth of value from 1 to 20 features. It does not change back in 20 features. The highest is **0.9878**. ACC values of this method is very high.

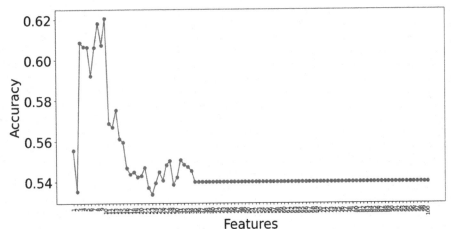

(a) Average ACC on Line Chart.

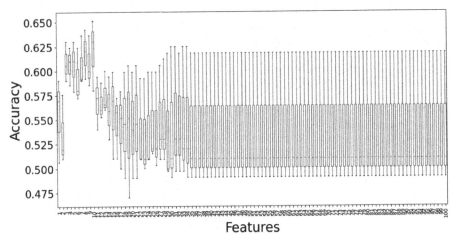

(b) The distribution and skewness of ACC on Boxplot Chart.

Fig. 3. Predict using SVM on CRC dataset (100 features) (Color figure online).

We can see that from Fig. 3 prediction by Support Vector Machine. First, Line chart Fig. 3a alteration is suddenly for the first 30 features and the highest ACC is **0.6205**. Second, Boxplot Fig. 3b also changes the same. Both graphs, after that values are virtually equal and the highest is **0.6513**. When you look at the two figures above you can clearly see the range of values and distribution of the ACC.

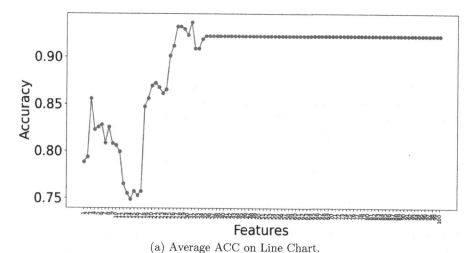

(a) Average ACC on Line Chart.

(b) The distribution and skewness of ACC on Boxplot Chart.

Fig. 4. Predict using SVM on IBD dataset (100 features) (Color figure online).

Both Figs. 3 and 4 are the same prediction method, but in Fig. 4 executes on IBD. It is also depicted in 2 graphs. The Line 4a and Boxplot 4b have an upward and reduction from 1 to 35 features. After in 35 features remain stable. The highest value of Line and Boxplot graph are 0.9365 and 0.9697 respectively. From the figure above it has been shown that this method also gives high results, but it is compared with RF it is less effective.

Values expressed in eight figures also show that ACC has a strong variation in the range from 1 to 35 features. When it is from 36 features onwards, most values did not change. And so, we only need to use about 35 features to get the desired results. This saves time and increases work efficiency see Table 3.

Table 3. Results in ACC of predictions on CRC datasets using Support Vector Machine (SVM).

	Features										The work in [27]
	1	2	3	4	5	6	7	8	9	10	
Yu	0.5697	0.5758	0.6303	0.6242	0.6303	0.6242	0.6364	**0.6424**	0.6364	0.6303	0.5534
Zeller	0.5066	0.5197	0.6053	0.5855	0.5789	0.5724	0.5921	0.5921	0.5855	**0.6513**	0.6711
Vogtmann	0.5900	0.5100	0.5900	0.6100	0.6100	0.5800	0.5900	**0.6200**	0.6000	0.5800	0.4905
Average	0.5554	0.5352	0.6085	0.6066	0.6064	0.5922	0.6062	0.6182	0.6073	**0.6205**	0.5717

Table 4. Results in ACC of predictions on IBD datasets using SVM and the work in [28].

	Features										The work in [28]
	1	2	3	4	5	6	7	8	9	10	
iCDr	0.7835	0.7938	**0.8351**	0.8041	0.8041	0.8041	0.7835	0.8041	0.8144	0.8144	0.843
UCf	0.7468	0.7848	**0.8481**	0.8101	0.8101	0.8228	0.7848	0.8101	0.7595	0.7595	0.902
iCDf	0.8049	0.7927	**0.8780**	0.8415	0.8537	0.8537	0.8415	0.8537	0.8293	0.8293	0.919
CDr	0.8000	0.8000	**0.8174**	0.8000	0.7913	0.7913	0.7739	0.7913	0.8000	0.8000	0.806
CDf	0.8061	0.7959	**0.8980**	0.8571	0.8673	0.8673	0.8571	0.8673	0.8367	0.8265	0.870
Average	0.7883	0.7934	**0.8553**	0.8226	0.8253	0.8278	0.8082	0.8253	0.8080	0.8059	0.868

The results confirm of Table 4, it is clear that we get ACC result when using Support Vector Machine to predict. It includes two types of values such as is the component ACC and the average ACC. On CRC without number of features, the highest ACC values in the dataset (Yu, Zeller, Vogtmann) are 0.6424, 0.6513, and 0.6200 respectively. The average is 0.6205. With IBD, the highest ACC values in the dataset (iCDr, UCf, iCDf, CDr, CDf) are 0.8351, 0.8481, 0.8780, 0.8174 and 0.8980 respectively. The average is 0.8553.

The two tables give information ACC results when using two methods to predict on two datasets. On CRC, when using RF and SVM, the results between the units with ACC are not much different. In addition, when we use two methods on IBD, the results varies enormously. RF included (iCDr: 0.9278, UCf: 0.9367, iCDf: 0.9634, CDr: 0.9130, CDf: 0.9592 and Average: 0.9383). In contrast, the SVM is (iCDr: 0.8351, UCf: 0.8481, iCDf: 0.8780, CDr: 0.8174, CDf: 0.8980 and Average: 0.8553). Therefore, we demomstrate that Random Forest is more effective than SVM.

In comparison with the works in [27] and [28], we obtained promising results. For CRC datasets, the selected features outperform the whole set of features in [27] for Yu and Vogtmann. With the top 10 features, the average performance is also better with 0.6205 comparing to 0.5717. For IBD datasets, we also achieved better results on CDf and CDr with only 3 features comparing to the whole set with more than 200 features with the work in [28].

From these results we can declare that select features using Random Forests and predict results using Random Forest can generate a high precision model at 35 features. Therefore, we can detect diseases easier with high precision.

To demonstrate the experimental results of this study are realistic, we have provided associated material at GitHub https://github.com/fptuni/random-forests.

5 Conclusion

Our research focuses on how to make good predictions for swiftly disease diagnosis and main aim is to help patients get better treated with early detection illness. As the predicted result sets have shown, we demonstrated high efficiency and accuracy when applying the Random Forest with two missions. First, It selects feature on Metagenomic Data provides some important features closely associated sickness. Second, Prediction brings the high accuracy desired. Moreover, this approach can reduce the number of features to enhance predictive performance while still delivering highly accurate results. This is the premise that helps medicine to develop quickly for research and treatment with technology application.

References

1. Handelsman, J., Rondon, M.R., Brady, S.F., Clardy, J., Goodman, R.M.: Molecular biological access to the chemistry of unknown soil microbes: a new frontier for natural products. Chem. Biol. **5**(10), 245–249 (1998)
2. National Library of Medicine: What is DNA? https://medlineplus.gov/genetics/understanding/basics/dna/
3. Handelsman, J.: Metagenomics: application of genomics to uncultured microorganisms. Microbiol. Mol. Biol. Rev. **68**(4), 669–685 (2004)
4. Sleator, R.D., Shortall, C., Hill, C.: Metagenomics. Lett. Appl. Microbiol. **47**(5), 361–366 (2008)
5. Amann, R.I., et al.: Phylogenetic identification and in situ detection of individual microbial cells without cultivation. Microbiol. Rev. **59**(1), 143–169 (1995)
6. Virgin, H.W., Todd, J.A.: Metagenomics and personalized medicine. Cell **147**(1), 44–56 (2011)
7. Ditzler, G., Polikar, R., Rosen, G.: Multi-layer and recursive neural networks for metagenomic classification. IEEE Trans. Nanobiosci. **14**(6), 608–616 (2015)
8. Soueidan, H., Nikolski, M.: Machine learning for metagenomics: methods and tools. Metagenomics **1** (2017)
9. El Naqa, I., Murphy, M.J.: What is machine learning? In: El Naqa, I., Li, R., Murphy, M.J. (eds.) Machine Learning in Radiation Oncology, pp. 3–11. Springer, Cham (2015). https://doi.org/10.1007/978-3-319-18305-3_1
10. Endo, K., Shiga, H., Kinouchi, Y., Shimosegawa, T.: Inflammatory bowel disease: IBD. Rinsho byori. Japan. J. Clin. Pathol. **57**(6), 527–532 (2009)
11. Ochsenkühn, T., Sackmann, M., Göke, B.: Inflammatory bowel diseases (IBD) - critical discussion of etiology, pathogenesis, diagnostics, and therapy. Der Radiologe. **43**(1), 1–8 (2003). https://doi.org/10.1007/s00117-002-0844-9
12. Dyson, J.K., Rutter, M.D.: Colorectal cancer in inflammatory bowel disease: what is the real magnitude of the risk? World J Gastroenterol. **18**(29), 3839–3848 (2012). https://doi.org/10.3748/wjg.v18.i29.3839. PMID: 22876036; PMCID: PMC3413056

13. Sleator, R., Shortall, C.; Hill, C. (2008, October 01). Metagenomics. Retrieved from https://sfamjournals.onlinelibrary.wiley.com, https://doi.org/10.1111/j.1472-765X.2008.02444.x

14. Nguyen, T.H., Zucker, J.: Enhancing metagenome-based disease prediction by unsupervised binning approaches. In: 2019 11th International Conference on Knowledge and Systems Engineering (KSE), Da Nang, Vietnam, pp. 1–5 (2019). https://doi.org/10.1109/KSE.2019.8919295

15. Nguyen, T.H., et al.: Disease classification in metagenomics with 2D embeddings and deep learning. In: The Annual French Conference in Machine Learning (CAp 2018). France, Rouen (June 2018). https://arxiv.org/abs/1806.09046

16. Nguyen, T.H., Nguyen, T.-N.: Disease prediction using metagenomic data visualizations based on manifold learning and convolutional neural network. In: Dang, T.K., Küng, J., Takizawa, M., Bui, S.H. (eds.) FDSE 2019. LNCS, vol. 11814, pp. 117–131. Springer, Cham (2019). https://doi.org/10.1007/978-3-030-35653-8_9

17. Nguyen, T., Chevaleyre, Y., Prifti, E., Sokolovska, N.; Zucker, J.: Deep learning for metagenomic data: using 2D embeddings and convolutional neural networks (2017, December 01). Retrieved November (2020) from https://arxiv.org/abs/1712.00244

18. Lladó Fernández, S., Větrovský, T., Baldrian, P.: The concept of operational taxonomic units revisited: genomes of bacteria that are regarded as closely related are often highly dissimilar. Folia Microbiol. **64**(1), 19–23 (2018). https://doi.org/10.1007/s12223-018-0627-y

19. What Is Colorectal Cancer?: How Does Colorectal Cancer Start? (n.d.). Retrieved from https://www.cancer.org/cancer/colon-rectal-cancer/about/what-is-colorectal-cancer.html

20. Segal, M.R.: Machine learning benchmarks and random forest regression. UCSF: center for bioinformatics and molecular biostatistics (2004). Retrieved from https://escholarship.org/uc/item/35x3v9t4

21. Genuer, R., Poggi, J.-M.: Random forests. In: Random Forests with R. UR, pp. 33–55. Springer, Cham (2020). https://doi.org/10.1007/978-3-030-56485-8_3

22. Vishwanathan, S.V.M., Murty, M.N.: SSVM: a simple SVM algorithm. In: Proceedings of the 2002 International Joint Conference on Neural Networks. IJCNN 2002. vol. 3, pp. 2393–2398, (Cat. No.02CH37290), Honolulu, HI, USA (2002). https://doi.org/10.1109/IJCNN.2002.1007516

23. Auria, L., Moro, R.A.: Support Vector Machines (SVM) as a technique for solvency analysis (August 1, 2008). DIW Berlin Discussion Paper No. 811, Available at SSRN: https://ssrn.com/abstract=1424949, https://doi.org/10.2139/ssrn.1424949

24. Sokol, H., Leducq, V., Aschard, H., et al.: Gut **66**, 1039–1048 (2017)

25. Dai, Z., Coker, O.O., Nakatsu, G., et al.: Multi-cohort analysis of colorectal cancer metagenome identified altered bacteria across populations and universal bacterial markers. Microbiome 6, 70 (2018). https://doi.org/10.1186/s40168-018-0451-2

26. Fioravanti, D., et al.: Phylogenetic convolutional neural net-works in metagenomics. BMC Bioinfor. **19**(2), 1–13 (2018)

27. Thanh-Hai, N., Thai-Nghe, N.: Diagnosis approaches for colorectal cancer using manifold learning and deep learning. SN Comput. Sci. **1**(5), 1–15 (2020). https://doi.org/10.1007/s42979-020-00297-7

28. Phan, N.Y.K., Nguyen, H.T.: Inflammatory bowel disease classification improvement with metagenomic data binning using mean-shift clustering. In: Dang, T.K., Küng, J., Takizawa, M., Chung, T.M. (eds.) FDSE 2020. CCIS, vol. 1306, pp. 294–308. Springer, Singapore (2020). https://doi.org/10.1007/978-981-33-4370-2_21

Predicting the Default Borrowers in P2P Platform Using Machine Learning Models

Li-Hua Li, Alok Kumar Sharma(✉), Ramli Ahmad, and Rung-Ching Chen

Department of Information Management, Chaoyang University of Technology, Wufeng, Taichung City 41349, Taiwan

{lhli,crching}@cyut.edu.tw, {s10814908,s10814904}@gm.cyut.edu.tw

Abstract. The online P2P platform's major advantage is that people can borrow or lend money free of intermediary interference. Prediction of the credit risk by the platform should ensure the borrowed money's repayment. This research used Random Forest (RF) in comparison with other machine learning (ML) techniques like Logistic Regression, K-Nearest Neighbor, and Multi-Layer Perception to predict the default borrowers. Lending Club's dataset is utilized for training and analyzing ML models. Statistical measures such as accuracy, recall, precision, F1-score, and the ROC curve are used to compare the data obtained in this study. The results were in accordance with Logistic Regression with the highest precision of 0.95 and RF with the highest AUC of 0.94. This study provides an overall understanding of different models and their prediction of default borrowers. Comparison of these models helps us to identify the most suitable model for the P2P platform.

Keywords: P2P lending · Credit risk assessment · Random Forest · Multi-Layer Perceptron · K-Nearest Neighbor · Logistic Regression

1 Introduction

Peer-to-peer (P2P) lending is a platform where money can be borrowed and lent directly between two individuals. This lending platform helps to bypass the intermediation of financial institutions, including banks. It strengthens the collaborative economy by providing an easily accessible and free platform for individuals who are often excluded due to poor financial status. The electronic lending platforms such as Lending Club (https://www.lendingclub.com/), Prosper (https://www.prosper.com/) and Kiva (https://www.kiva.org/) playing a pivotal role in providing free loaning services between borrowers and lenders [1]. This emerging platform offers lenders a reliable and cheaper alternative to connect with the borrowers. The loan applications are submitted electronically, which are accessible to the lenders. Based on the borrower information, the lenders make their decisions [2]. Lending Club, in 2018, processed a 1.4 million number of loans worth 20 billion USD. Lending Club, in 2018, processed a 1.4 million number of loans worth 20 billion USD. Lending Club offers faster and easier online/mobile-based loan processing options at a cheaper interest rate. However, P2P lending faces challenges in its development, such as asymmetric information and improper risk handling methods.

A. Solanki et al. (Eds.): AIS2C2 2021, CCIS 1434, pp. 267–281, 2021.
https://doi.org/10.1007/978-3-030-82322-1_20

Such issues bring in the possibility of defaults in P2P lending. These defaults can be unfavourable for the lenders and the development of P2P lending platforms [3]. Risk assessment is an important aspect that could be done using loan evaluation tools. The purpose of credit risk evaluation is to help investors making profitable decisions.

Recently, several researchers [4–9] have developed various machine learning-based loan evaluation models for assessing credit risk. The Random Forest (RF) model is one model which uses a generic algorithm. It has an excellent performance record in evaluating loan at P2P lending platforms [3]. Another model, known as Multi-Stage Ensemble Learning, has been developed to classify the borrower credit, which is also useful to P2P lending services [4]. Since not enough research focuses on RF model in light of P2P landing, this research attempts to apply the RF model approach. It compares multiple classification models for examining the P2P lending club credit risk and increasing the assessment accuracy. The motivation behind this study is that the lender lends their money to borrowers through this platform. Therefore, this is better if the lender could be known that the borrower will be a defaulter or not. This prediction can be beneficial for investors. In this research, we have selected ten features as input for predict default borrower. This study employs Random Forest (RF) and compares it with other machine learning models like Logistic Regression (LR), K-Nearest Neighbor (KNN) and Multi-Layer Perceptron (MLP). This research aims to predict default borrower in the P2P lending platform and experiment with which machine learning model (mentioned above) is suitable for default borrower prediction.

The remainder of this paper is prepared as follows. Section 2 highlights related background research. The research methodology is described in Sect. 3, followed by results and analysis in Sect. 4. Further, this research is concluded in Sect. 5 and discusses the future directions.

2   Background Studies

2.1   Overview of P2P Lending

Peer-to-peer (P2P) lending is a rapidly growing form of online micro-financing that can alternate traditional credit financing. The P2P lending model utilizes internet-based platforms in financing (lend/borrow) devoid of financial intermediation [5]. The step process of P2P lending starts with the submission of loan request by the borrowers, which includes information such as purpose and amount of loan requested, the possible interest that can be paid, and borrower's financial & personal details, which will be processed through the lending platform. The loan requests will then be listed in the market for bidding by potential investors, who, based on available details on the demographic characteristics, the financial strength of the borrower, effort indicators, amount of loan requested with interest rate and duration given by the borrower in the lending platform, select the loan applications to invest. The borrower will then be provided with the loan's requested sum entirely when the lender fully funds the loan, or enough bids were obtained on the request [5]. Finally, the process ends with timely lending of the loan amount to the borrowers without collateral, leaving the lenders offered with high returns for their investments.

P2P lending showed steady growth since its emergence in developed countries like Britain in 2005 when Zopa was developed as the first network lending website (https://www.zopa.com/), followed by Prosper's development in 2007, which was America's first P2P lending website (https://www.prosper.com/). Since then, P2P platforms have been showing influence worldwide and emerged as a significant part of the financing market [6]. Previous research has also tried to perform a decision tree-based classification technique for P2P application traffic [10].

Among several factors that influence the investors' decision, trust between the borrower and lender holds prime significance. Trust between lenders and borrowers is a critical decision-making factor. Constant interaction with the exchange of messages between borrowers and lenders, including simple factors like borrowers' picture, impacts building trust [5]. Besides, when selecting a loan offer, the lenders show herding behavior, as they prefer to invest in loans with more bids. The major attraction of P2P leading is the absence of intermediaries and collateral, which makes this internet-based lending platform to have the edge over securing loans from traditional banks. The loan requested is generally for small to medium range investments, with a less than or equal to three years as standard payback period. Apart from the convenient processing, both lenders and borrowers are mutually benefitted through the P2P lending platform.

2.2 Random Forest (RF)

Random Forest (RF) [11] is a decision tree based ensemble learning algorithm that can be applied for both regression and classification tasks, while its implementation is straightforward. Random Forest obtained original sample data by bootstrapping. Each tree provides a classification and the tree with the most votes among all the trees will be selected by forest. The parameter m denotes the number of decision trees for determining the degree of randomness. Generally, when it was assumed when borrower possesses d attributes, set $m = log_2 d$. Through the Gini index, the CART classification [11] creates the split points. Considering n classes, p_i is the probability that an object belongs to class i, then the Gini index [11] can be derived as follow.

$$Gini(D) = 1 - \sum_{i=1}^{n} p_i^2 \tag{1}$$

Single decision trees follow a decision-making method that is very similar to the humans' decision-making process, which possesses a chain of simple rules. However, decision trees' accuracy in the generalization of unpredictable samples, whereas Random Forest is a useful prediction tool.

2.3 Logistic Regression (LR)

One of the most widely used classification model for machine leaning is Logistic Regression (LR) [8]. LR is an economical methodology for computational costs. LR calculates the probability of a sample belonging to 0 or 1. LR has been utilized as a credit score model in previous research as [8]. As shown below in Eq. (2), the Sigmoid function is used for LR to converge (see Fig. 1):

$$h = \frac{1}{1 + e^{-x}} \tag{2}$$

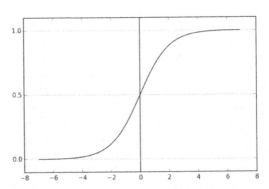

Fig. 1. Sigmoid function for LR.

For Eq. (2), set x as the borrower's function vector, and h is the set of values corresponding to each feature. Logistic Regression (LR) [12] is a statistical method, which derives "an equation that predicts an outcome for a binary variable (Y) from one or more response variables (X) and this method is similar to linear regression." However, the response variables can be categorical or continuous since, unlike Linear Regression, the model does not strictly require continuous data. LR relies on the log odds ratio rather than probabilities to predict group membership. An iterative maximum probability approach is more fitting than a minimum square to match the final model. Therefore, by using LR, researchers obtain more freedom because it is more acceptable for non-normally distributed data or when the samples have unequal covariance matrices. Logistic Regression assumes independence between variables, which in datasets is not always met. However, a common phenomenon is the system's applicability (and how well it performs, e.g., the classification error) exceeding statistical assumptions. One Inability to generate probabilities of typicality (useful for forensic casework) is a single drawback of LR, but nonparametric methods such as ranked probabilities and ranked inter-individual similarity measures may substitute these values.

2.4 K-Nearest Neighbor (KNN)

K-Nearest Neighbors [7] technique can be used for both regression and classification tasks. To predict the data point and its category, KNN explores the labels of a chosen number of data points surrounding a target data point. KNN is well-known for its simple and effective algorithm and is considered one of the widely used machine learning algorithms.

KNN [13] can be defined as a supervised learning algorithm since the examples in the dataset must have labels assigned to them, i.e., the class of each data must be known. The critical point to note is that KNN is a non-parametric algorithm [13], so no assumptions are made about the dataset when the model is used, and the model is entirely built from the data given. Further, datasets were not split into training and testing sets and makes no generalizations between them. So, all the training data were also used for predictions with this model.

2.5 Multi-Layer Perceptron (MLP)

MLP [9] is a relatively simple form of neural network (see Fig. 2 for reference) due to its unidirectional flow of information from input nodes to output nodes and is referred as front propagation only. Backpropagation, on the other hand, is a training algorithm in which errors are determined after forward values are fed and then backpropagated to earlier layers. To put it another way, backpropagating signals from nodes occurs after forwarding propagation, which is a part of the backpropagation algorithm.

Also, the network needs not possess a hidden layer that makes MLP a part of artificial neural networks called feed-forward neural networks.

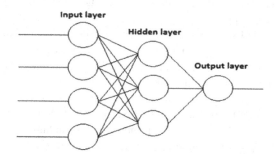

Fig. 2. MLP with one hidden layer.

2.6 Recursive Feature Elimination (RFE)

RFE [14] is an algorithm for selecting wrapper-type features, where an altered machine learning algorithm is used in the center of the process, wrapped by RFE, to help select features. Unlike filter-based feature selection, the RFE algorithm does not rate each feature and select the features that have the largest (or smallest) score. RFE is a standard algorithm for feature selection that often internally uses filter-based feature selection.

RFE [9] works by searching for a subset of features among all features in the training dataset were explored by the RFE algorithm, thus successfully removing features until the target number was stable. This is accomplished by first fitting the known machine learning algorithm used in the model's core, ranking critical features, and discarding the least significant one, finally re-fitting the model. This procedure is repeated until a defined number of characteristics remain.

2.7 Previous Research in P2P Lending

Many researchers (as shown in Table 1) have used machine learning and other techniques to analyze the credit risk assessment. Previous research (see Table 1) has also used algorithms such as Logistic Regression (LR) [12, 15–19], Neural Network (NN) [5, 15–17, 20], Random Forest (RF) [12, 18] and Decision Tree (DT) [8, 18, 21, 22]. Table 1 summarizes these methodologies and dataset used by various researchers about P2P lending analysis.

As shown in Table 1, researchers [8, 12, 16] employed LR model to evaluate the credit risk for P2P lending using dataset from Lending Club. Other researchers [17, 18, 25] have also utilized LR Model for credit risk analysis using other lending platform like Yooli, Eloan and Paipai. Furthermore, [19, 20, 23] have used deep learning for analyzing credit risk of P2P lending using dataset from Lending Club. Different machine learning models like Naïve Bayes, Decision Tree and Gradient boosting decision trees are employed by previous research for credit risk prediction based on various P2P platform. Byanjankar et al. [5] applied Artificial Neural Networks (ANNs) for finding the solution using Bondora dataset.

Previous research highly applied machine learning models for P2P lending, credit risk analysis, and default prediction (Table 1). Previous research [26] has applied linear regression, ARIMA, and support vector regression to predict the resource workload in the cloud environment. All researchers have applied various machine learning models to predict default borrowers. However, research on RF in a similar context has not been well explored. Chen et al. [12] applied LR for measuring credit risk on P2P lending and compared the results with RF. Jian et al. [18] compared various models, including RF for default loan prediction using description text from the Eloan P2P lending platform. Approaches were proposed to improve their models to obtain higher accuracy and AUC (Area Under the ROC Curve) value. This research further elaborates the RF model approach to predict the borrower's default situation for P2P lending and comparing the same with other machine learning models such as LR, KNN and MLP.

Table 1. Related studies in chronological order for credit risk analysis.

Year	Author	Dataset	Data	Method
2020	Zanin [15]	Lending Club	612,745	Logistic Regression
2020	Song [21]	Lending Club	162,570	Gradient Boosting Decision Trees
2019	Kim and Cho [20]	Lending Club	855,502	Convolutional Neural Networks, Deep Learning
2019	Wang et al. [19]	P2P Lending Enterprise in China	100,000	Deep Learning
2019	Tan et al. [23]	Lending Club	132,1864	Deep Learning

(continued)

Table 1. (*continued*)

Year	Author	Dataset	Data	Method
2019	Chen et al. [12]	Lending Club	117,790	Logistic Regression, Random Forest
2018	Shuai Ding et al. [16]	P2P Lending Enterprise in China	80,000	Deep Neural Network, XGBoost and Logistic Regression
2018	Kvamme et al. [24]	Norwegian Financial Service Group	20,989	Deep Learning
2017	Kim and Cho [22]	Lending Club	332,844	Decision Tree
2017	Lin et al. [17]	Yooli	48,784	Logistic Regression
2017	Jiang et al. [18]	Eloan	39,538	Logistic Regression, Naïve Bayes, Random Forest, Support Vector Machine
2017	Zhang et al. [25]	Paipai	193,614	Logistic Regression
2016	Serrano-Cincad and Gutiérrez-Nieto [8]	Lending Club	40,907	Linear Regression, Decision Tree
2015	Byanjankar et al. [5]	Bondora	16,037	Artificial Neural Networks

To improve the prediction performance, this research will propose the improved process flow of RF and compare it with other machine learning models as proposed by the previous researchers. This research will apply Recursive Feature Elimination (RFE) in the feature's selection process. To prove if proposed RF model is applicable and has better performance, this research also utilizes the dataset from Lending Club and compares the model with other machine learning models such as LR, KNN and MLP.

3 Methodology

In this research, the study's main goal is to predict the defaulting borrower in the P2P platform using machine learning models such as RF, LR, KNN and MLP. This study also compared these machines learning classification methods and evaluate which model is better for prediction. The process flow (see Fig. 3) of this research is as follow.

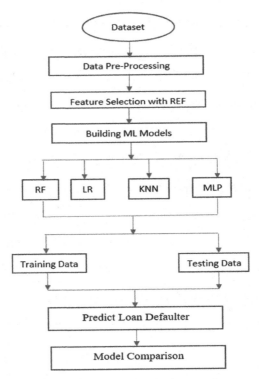

Fig. 3. Flow diagram of this research

3.1 Dataset and Pre-processing

The Borrower data of LendingClub, i.e., Borrower's dataset, is collected from the Kaggle website (https://www.kaggle.com/). Each day the P2P lending platform generates a considerable amount of data. This research extracted data using the period from Jan 2018 to Dec 2018 of Lending Club data. Our data includes approximately 411K borrower records, containing 145 features and some features have a null value in the dataset. Firstly, authors removed all the empty features and some other irrelevant features also removed, which is not essential for our objective data cleaning. Since the Lending Club company did not reveal borrowers' personal information, there are several empty and meaningless features in the dataset, which are further removed from the dataset.

After the data cleaning process, this study applies the dataset's correlation matrix and refer to the data dictionary of Lending Club. Twenty features related to our prediction task are selected for modelling building and training. The result of data preprocessing is to select highly correlated features and to clean the dataset into the size of (11000, 20) where 11000 means the number of data records and 20 means the total number of features.

3.2 Feature Selection

For finding the best features for our study. This research has applied the Recursive Feature Elimination (RFE) LR model [9, 14], to obtain the most important ten features, as shown in Table 2. Additionally, these ten features are used for the proposed RF models.

Table 2. Selected features for our model

S. No.	Features	Description
1	funded_amnt	Total loan amount invested
2	annual_inc	Information regarding the borrower's yearly income provided during registration
3	emp_length	Duration of employment (Years). Values lie between 0 and 10 with 0 denotes 1 year or less and 10 denote duration above ten years
4	mort_acc	Information of Number of mortgage accounts
5	last_pymnt_amnt	Final amount received in total
6	int_rate	Rate of interest
7	mo_sin_old_rev_tl_open	Time (in months) since the opening of a revolving account (oldest)
8	acc_open_past_24mths	Trades opened previously for 24 months
9	avg_cur_bal	Existing balance of all types of accounts and its average
10	Num_sats	The number of accounts that are satisfactory

The variable predicted in our dataset is 'loan_status', which provides the loan status as three kinds of values – "Fully Paid," "Default," and "Charged Off." The meaning is explained as below.

Fully Paid: Loan has been paid back completely.
Default: Loan has not been paid back 121 days or more.
Charged Off: A loan on which the presumption of further payments is no longer realistic.

Since our goal in this study is to predict whether or not a borrower will default on the loan, this research has considered the findings where the loan status is unless charged-off or fully paid. Data of "Charged Off" and "Default" is set as 1, and the data of "Fully Paid" is set as 0, where "1" means the borrower as a defaulter.

3.3 Building Machine Learning Models

Our classification issue is to predict whether or not the borrower will default. This research used RF as a classification model. This study used the Gini index to evaluate if the defaulting borrower is predicted, which has already been mentioned above. Based on our model training process, This study has found (see Fig. 4) "last_pymnt_amnt" feature has a significant weighting of more than 30%. In each of its decision trees, RF chooses a subset of characteristics, thus reducing the model's bias (due to the high significance of one feature). The final output would be the output mode of all of its decision trees that have better results than decision trees (which can overfit). That's why this research have chosen to start our classification with RF.

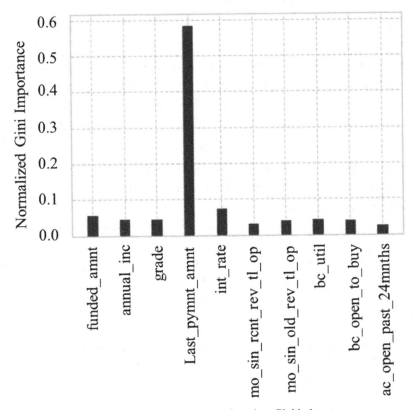

Fig. 4. Feature importance based on Gini index

To perform RF, LR, KNN, and MLP algorithm experiments, this research used Anaconda (https://www.anaconda.com/products/individual) and Jupiter notebook for coding and model building. The study deployed the program using the python Sklearn library. Further, the ROC curve and the confusion matrix are calculated. Confusion matrix consists of True Positive *(TP)*, True Negative *(TN)*, False Positive *(FP)* and False Negative *(FN)* [27].

$$accuracy = \frac{TP + TN}{TP + FP + FN + TN} \qquad (3)$$

$$precision = \frac{TP}{TP + FP} \qquad (4)$$

$$recall = \frac{TP}{TP + FN} \qquad (5)$$

$$F1\ score = \frac{2 * recall * precision}{recall + precision} \qquad (6)$$

The experiment outcomes of each model by using the matrix of Eqs. (3)–(6) are displayed as in Table 4.

3.4 Training and Testing Data

After preprocessing for this study, 11000 datasets for the experiment were used and dataset was split into two parts: Training Dataset and Testing dataset. Authors kept 20% of data as testing data and 80% for training (shown in Table 3). The dataset has 5500 fully paid and 5500 default & charged-off loan status data. Furthermore, this study chooses random loan status data for Training and Testing. Training dataset has used to train our model and Testing data have used to test the experiment.

Table 3. Data split summary

Samples	No. of dataset	Percentage (%)
Training	8,800	80
Testing	2,200	20
Total	11,000	100

4 Experiment and Result Analysis

The dataset of LendingClub was collected from from Jan. 2018 to Dec. 2018 from the Kaggle (https://www.kaggle.com/). After data cleaning and pre-processing, as mentioned in Sect. 3.1, the study obtained 11,000 borrowers' loan data with 20 features at the beginning. After applying the RFE method, the best 10 features are obtained for the study (as mentioned in Table 2).

This research has performed all experiments on MacBook Pro with 2.4 GHz Intel Core i7 CPU, and 16 GB of memory and used Python (version 3.7.1) programming language for model building and data analysis.

Using the prepared data subset, this research trained four models, after which the models obtained the classification results in terms of the test data sets. The ROC curves of computational outcomes in these models are shown in Fig. 5. Table 5 shows that the RF model has the highest AUC: 0.94, the LR & MLP model has a similar AUC: 0.93, and KNN has the minimum AUC: 0.88.

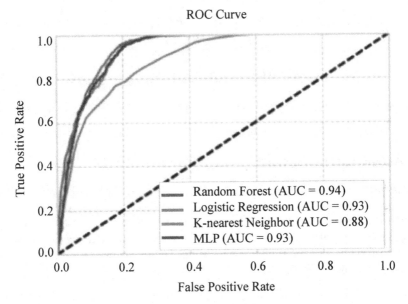

Fig. 5. Comparison of ROC curve

Table 4. The prediction results of various models

	Accuracy	Precision	Recall	f1-score
RF	**0.87**	0.92	**0.82**	**0.87**
LR	**0.87**	**0.95**	0.79	0.86
KNN	0.79	0.83	0.73	0.78
MLP	**0.87**	0.93	0.80	0.86

Table 5. Result of AUC with various models

Model	AUC
RF	**0.94**
LR	0.93
KNN	0.88
MLP	0.93

This study also calculated the Accuracy, Precision, Recall and f1-score as shown in Table 4, and it was identified that RF, LR & MLP have similar accuracy of 0.87. LR have the highest precision score of 0.95 and KNN have the lowest of 0.83. RF have the highest recall score of 0.82, and KNN has the lowest recall score of 0.73. Furthermore, for the f1-score, RF had the highest score of 0.87, and KNN had the lowest score of 0.78. The higher performance of RF can be attributed to its higher AUC value in the ROC curve. Since higher AUC values attribute to the better classification models. Although logistic regression has higher precision among other models, RF still dominates the results if evaluated based on AUC.

5 Conclusion and Future Scope

Lending Club dataset utilized in this research is considered as a sufficient database for analyzing borrowers' P2P lending data. The period for our analysis started from Jan 2018 to Dec 2018. This study applied Random Forest (RF) for predicting default borrowers and our method has compared with other machine learning models, including Logistic Regression (LR), K-nearest Neighbor (KNN), and Multilayer Perceptron (MLP). The AUC results show that RF is the best model with AUC: 0.94 if compared with LR's ACU: 0.93, KNN's AUC: 0.88 and MLP's AUC: 0.93.

While comparing the statistical parameters of different testing models in this study, RF is observed to be in sync with other models such as LR, and MLP. However, while considering ROC, the study identifies that Random Forest performs better than other selected models. Higher AUC is measured as the indicator of better model performance. RF is considered high-performance metrics in terms of accuracy, recall, and f1-score, but the time RF takes to run the codes is also of minimum duration.

This research identifies that "last_payment_amount" is an essential feature for default prediction from our research experiments results. Therefore, authors will work on this feature for further study in the future. Future research should also consider more technologies, such as Deep Learning, the ARIMA model, Convolutional Neural Network (CNN), etc. Integration of fine-tuned data preprocessing for multi-stage prediction of credit risk can also be a promising approach for the default prediction of P2P lending.

References

1. Serrano-Cinca, C., Gutiérrez-Nieto, B., López-Palacios, L.: Determinants of default in P2P lending. PLoS ONE **10**, 1–22 (2015)
2. Shen, F., Wang, R., Shen, Y.: A cost-sensitive logistic regression credit scoring model based on multi-objective optimization approach. Technol. Econ. Dev. Econ. **26**, 405–429 (2020)
3. Ye, X., Dong, L., Ma, D.: Loan evaluation in P2P lending based on Random Forest optimized by genetic algorithm with profit score. Electron. Commer. Res. Appl. **32**, 23–36 (2018)
4. Chen, S., Wang, Q., Liu, S.: Credit risk prediction in peer-to-peer lending with ensemble learning framework. In: Proceedings of the 31st Chinese Control Decision Conference (CCDC 2019), pp. 4373–4377 (2019)
5. Byanjankar, A., Heikkila, M., Mezei, J.: Predicting credit risk in peer-to-peer lending: a neural network approach. In: Proceedings of the IEEE Symposium Series on Computational Intelligence (SSCI 2015), pp. 719–725 (2015)
6. Zhou, G., Zhang, Y., Luo, S.: P2P network lending, loss given default and credit risks. Sustainability **10**, 1–15 (2018)

7. Nukala, S.D., Mishra, V.K., Nookala, G.K.M.: Modeling earthquake damage grade level prediction using machine learning and deep learning techniques. In: Sharma, N., Chakrabarti, A., Balas, V.E., Martinovic, J. (eds.) Data Management, Analytics and Innovation. AISC, vol. 1175, pp. 421–433. Springer, Singapore (2021). https://doi.org/10.1007/978-981-15-5619-7_30

8. Serrano-Cinca, C., Gutiérrez-Nieto, B.: The use of profit scoring as an alternative to credit scoring systems in peer-to-peer (P2P) lending. Decis. Supp. Syst. **89**, 113–122 (2016)

9. Duan, J.: Financial system modeling using deep neural networks (DNNs) for effective risk assessment and prediction. J. Franklin Inst. **356**, 4716–4731 (2019)

10. Bhatia, M., Sharma, V., Singh, P., Masud, M.: Multi-level P2P traffic classification using heuristic and statistical-based techniques: a hybrid approach. Symmetry (Basel) **12**, 2117 (2020)

11. Niu, B., Ren, J., Li, X.: Credit scoring using machine learning by combing social network information: evidence from peer-to-peer lending. Information **10**(12), 397 (2019)

12. Chen, S.-F., Chakraborty, G., Li, L.-H.: Feature selection on credit risk prediction for peer-to-peer lending. In: Kojima, K., Sakamoto, M., Mineshima, K., Satoh, K. (eds.) JSAI-isAI 2018. LNCS (LNAI), vol. 11717, pp. 5–18. Springer, Cham (2019). https://doi.org/10.1007/978-3-030-31605-1_1

13. Caplescu, R.D., Panaite, A.M., Pele, D.T., Strat, V.A.: Will they repay their debt? Identification of borrowers likely to be charged off. Manag. Mark. **15**, 393–409 (2020)

14. Li, X., Li, X., Liu, W., Wei, B., Xu, X.: A UAV-based framework for crop lodging assessment. Eur. J. Agron. **123**, 126201 (2021)

15. Zanin, L.: Combining multiple probability predictions in the presence of class imbalance to discriminate between potential bad and good borrowers in the peer-to-peer lending market. J. Behav. Exp. Financ. **25**, 100272 (2020)

16. Li, W., Ding, S., Chen, Y., Yang, S.: Heterogeneous ensemble for default prediction of peer-to-peer lending in China. IEEE Access **6**, 54396–54406 (2018)

17. Lin, X., Li, X., Zheng, Z.: Evaluating borrower's default risk in peer-to-peer lending: evidence from a lending platform in China. Appl. Econ. **49**, 3538–3545 (2017)

18. Jiang, C., Wang, Z., Wang, R., Ding, Y.: Loan default prediction by combining soft information extracted from descriptive text in online peer-to-peer lending. Ann. Oper. Res. **266**, 511–529 (2018)

19. Wang, C., Han, D., Liu, Q., Luo, S.: A deep learning approach for credit scoring of peer-to-peer lending using attention mechanism LSTM. IEEE Access **7**, 2161–2168 (2019)

20. Kim, J.Y., Cho, S.B.: Towards repayment prediction in peer-to-peer social lending using deep learning. Mathematics **7**(11), 1041 (2019)

21. Song, Y., Wang, Y., Ye, X., Wang, D., Yin, Y., Wang, Y.: Multi-view ensemble learning based on distance-to-model and adaptive clustering for imbalanced credit risk assessment in P2P lending. Inf. Sci. (NY) **525**, 182–204 (2020)

22. Orchard, J., Castricato, L.: Combating adversarial inputs using a predictive-estimator network. In: Liu, D., et al. (eds.) ICONIP 2017, Part II, LNCS 10635, pp. 118–125 (2017). https://doi.org/10.1007/978-3-319-70096-0_13

23. Tan, F., Hou, X., Zhang, J., Wei, Z., Yan, Z.: A deep learning approach to competing risks representation in peer-to-peer lending. IEEE Trans. Neural Netw. Learn. Syst. **30**, 1565–1574 (2019)

24. Kvamme, H., Sellereite, N., Aas, K., Sjursen, S.: Predicting mortgage default using convolutional neural networks. Expert Syst. Appl. **102**, 207–217 (2018)

25. Zhang, Y., Li, H., Hai, M., Li, J., Li, A.: Determinants of loan funded successful in online P2P lending. Procedia Comput. Sci. **122**, 896–901 (2017)

26. Singh, P., Gupta, P., Jyoti, K.: TASM: technocrat ARIMA and SVR model for workload prediction of web applications in cloud. Clust. Comput. **22**, 619–633 (2018)
27. Hindistan, Y.S., Aiyakogu, B.A., Rezaeinazhad, A.M., Korkmaz, H.E., Dag, H.: Alternative credit scoring and classification employing machine learning techniques on a big data platform. In: Proceedings of the 4th International Conference on Computer Science and Engineering (UBMK 2019), pp. 731–734 (2019)

Human Activity Recognition for Multi-label Classification in Smart Homes Using Ensemble Methods

John W. Kasubi[1,2](✉) [iD] and Manjaiah D. Huchaiah[1,2] [iD]

[1] Department of Computer Science, Mangalore University, Mangalore, India
[2] Local Government Training Institute, Dodoma, Tanzania

Abstract. Human Activity Recognition (HAR) plays a vital role in recognizing human activities in smart homes environment. HAR is a prominent research area of great significance due to its applicability in different fields such as security, elderly care, and healthcare, to name a few. The study focused on recognizing human activities of daily living using the ARAS dataset, and for this reason, Bagging and Adaboost ensemble methods were employed. The study involved five popular classification algorithms, Logistic Regression (LR), Random Forest (RF), k-nearest neighbours (KNN), Naive Bayes (NB), and Support Vector Machine (SVM), as a base of ensemble classifiers. The experimental results show that the Adaboost ensemble method outperformed compared to the Bagging ensemble Method in both House A and B through Random Forest (RF) classifier.

Keywords: Human activity recognition · Multi-class classification · Smart home · Ensemble method

1 Introduction

The HAR examines various activities of daily living (ADL) of human performed in the household components connected to sensors. HAR is used to mine meaningful information from raw data, which gives insight into human activities performed in the presence of sensors [1]. HAR is an exciting research topic for human behaviour monitoring that facilitates different applications in different fields such as surveillance, healthcare monitoring; energy monitoring; identification of behaviour trends; water resource management, and home situation assessment [2]. HAR plays a significant and essential task in smart homes for predicting potential behaviour performed by humans at home. Results may lead to security monitoring, healthcare, water, and energy management. With the help of HAR, smart homes keep human life safe by the assurance of security all the time, early disease detection, control of energy usage at homes, water management, etc., hence, quality life [3].

Smart Homes are the backbone of a Smart City, whereby multiple smart homes or buildings are connected through networks as a whole. The Smart Homes helps city's residents to live a quality life by praying an essential role in various fields such as healthcare due to early diseases detection; financial saving due to control and management of

© Springer Nature Switzerland AG 2021
A. Solanki et al. (Eds.): AIS2C2 2021, CCIS 1434, pp. 282–294, 2021.
https://doi.org/10.1007/978-3-030-82322-1_21

energy grids and the usage of water, security and surveillance issues, hence, quality life and sustainable development [4].

1.1 Ensemble Methods and Base Classifiers

The ensembles methods combine various algorithms and can be addressed in multiple ways. In this study, two approaches were used; Bagging and Adaboost and five base learner classifiers; LR, RF, KNN, NB, and SVM. The bagging method is a powerful and straightforward ensemble approach. The bagging technique combines the predictions from several classifiers to make predictions more precise than any individual model from a different random subset of the training dataset. The Adaboost method builds multiple models of the type using a sequential set of algorithms to create a robust model to improve the performance [5]. LR- It is easier to execute, interpret, and train very effectively, while RF- It is the most powerful machine learning algorithm and type of bagging ensemble method. It gives more accuracy and avoids overfitting [6]. KNN- is simpler to implement, does not make any assumptions, and is based on features' similarity. For better accuracy, it uses turning parameters to select the correct value of 'k'. At the same time, NB - is the set of probabilistic classifiers that assume that all the features used for classification are independent of each other. The outcomes of the model depend on the sets of independent variables [7]. SVM- classifies data into different classes using a hyperplane which divides the features into the best possible two categories using a kernel technique. The hyperplane is drawn based on the data point closed to it called support vectors with the target to get the best line that segregates the two classes; SVM is usually faster and more successful [8]. These ensemble methods and five base learner classifiers have been used in this study because are simple to implement. Also, compared to other approaches, which are even quicker, produce better performance, boost model prediction robustness, and avoid overfitting. This study used the ARAS dataset, which consists of 27 different types of activities conducted by multiple residents from both House A and B. The ARAS dataset is an imbalanced dataset with an unequal number of instances in each class, making it more challenging to analyze accurate activity identification. The ARAS dataset is a multiclass classification problem with more than two categories, hence, more challenging than a binary classification problem [9].

In this regard, the study focused on developing a predictive model using the ensemble approach to identify ADLs in smart homes using the ARAS dataset to create better predictions than any individual model. For this matter, seven algorithms were employed, and the particular outcomes were combined to form one best prediction than any specific model.

1.2 The Motivation and Contributions of This Paper

HAR plays a significant role in smart homes in different area such as security, healthcare, automation, water and electricity usage. The HAR helps diagnose disease in smart homes at the early stage, hence better health for smart homes residents. HAR helps to identify visitors trying to disturb homes for security issues, which can be reported to the nearest police station to take necessary measures. The HAR is used to reduce the financial expenses of energy and water usage. This paper includes the following contributions:

- The ensemble's suggested method increased the better HAR predictions in ADL for multiple residents in smart homes and produced better results than any single model that contributes.
- The ensemble methods improved the robustness of the prediction model for HAR by combining the individual classifier's performance compared to the previous study, as shown in Table 4.

This paper is organized as follows: Sect. 2; related works; Sect. 3; presents methodology of the study; Sect. 4; presents the experimental results and discussions; and finally, in Sect. 5; provides the conclusions and give suggestions for future work.

2 Related Research Works

This chapter section summarises previous relevant works examined concerning HAR for Multi-class Classification Dataset in Smart Home; the related research reviewed were as follows:

Elamvazuthi et al. [10] presented an ensemble approach on HAR using a wearable sensor to identify six ADL on the dataset collected from the UCI Machine Learning Repository. The study employed five types of ensemble classifiers: Adaboost, Bagging, END, Random subspace, and Rotation forest, using SVM and RF on the back end. The SVM performed best compared to RF with an accuracy of 99.22% and 97.91%, respectively. Ensemble classifier improves performance; the suggestion was given to employ different methods to improve performance.

Adama et al. [11] studied HAR using assistive robots to recognize human activities; in this regard, researchers employed an ensemble of three individual classifiers includes; RF, KNN, and SVM, KNN, on the CAS-60 dataset. The performance of the suggested approach was enhanced compared to the previous one of using an individual classifier. The researcher advised to include more classifiers to improve the performance and apply DNN to increase HAR.

Altuve et al. [12] presented research on classification for six ADL using bidirectional LSTM on UCI Machine Learning Repository dataset collected through a smartphone. The total outcomes achieved were 92.9% accuracy; this performance resulted from using the ensemble approach.

Padmaja et al. [13] introduced ERT classifier in HAR using two datasets, HAPT and HAR, with 12 and 6 activities, respectively. The outcome shows that the suggested method did better than the existing approaches, with an accuracy of 92.63% for the HAPT dataset and 94.16% for the HAR dataset.

WH Chen [14] used five popular classification techniques, SVM, NB, RL, RFB, and DT, to build a bagging classifier, focusing on activities recognition allied to ADLs based on binary sensor data. The results show that the suggested method performed best, indicating that combining the classifier improves model performance.

Zhenghua et al. [15] propose an ELM classifier for HAR using smartphone sensors to identify six activities using two datasets. The study employed ANN, SVM, ELM, RF, and LSTM for ensemble algorithms purposes. The experimental results point out that the proposed ensemble ELM approach outperformed and managed to obtain accuracies of 98:88% and 97:35% onto two datasets.

Irvine et al. [16] experimented on recognizing ADL activities using the homogeneous ensemble NN method in smart environments. The ensemble NN outperformed with an accuracy of 80.39% by showing the efficacy of the recommended suggested approach.

Jethanandani et al. [17] assessed KNN, NB, RL, and DT's performance in a multi-label classification dataset using the Majority Voting Ensemble algorithm method to detect ADL in the ARAS dataset. The experiments were done and obtained an accuracy of 88% for House A and 97% for House B. Suggestions were given to use different techniques to address the HAR problem.

Ni, Qin, et al. [18] used a wearable accelerometer dataset to recognize ADL in smart homes. Three algorithms were selected in the ensemble method: KNN, DT, and SVM. The suggested approach obtained an accuracy of 96.87%.

Saha et al. [19] presented a smartphone for HAR and used the ensemble method to address the collected dataset variance. It was observed that the proposed approach achieved an accuracy of 94% in recognition of ADL.

Bulbul et al. [20] experimented on HAR using three ensemble classification methods, boosting, bagging and stacking together with DT, SVM and KNN as base leaner classifiers. The authors achieved 94.4% accuracy by using DT, 99.4% through SVM and 97.5% from KNN. The experimental results obtained through AdaBoost was 97.4% accuracy, Bagging was 98.1% accuracy and Stacking 98.6% accuracy.

Subasi et al. [21] applied the Adaboost ensemble classifier for body sensors; this method achieved better outcomes by suing different classifier models to perform activity recognition better. The result showed that classifiers based on Adaboost perform better in HAR.

Tian et al. [22] presented the majority voting method in recognizing HAR using three base classifiers, SVM, KNN and DT. The proposed approach outperformed by getting the average recognition accuracy of 94.79%.

Wang et al. [23] used -based on HAR in better tracking, using Bagging and Adaboost methods and obtained 92.3% accuracy and 90.7% recall, which shows better performance in activity sensor recognition.

Xu, S. et al. [24] proposed a cascade ensemble learning public datasets of HAR based on the smartphone, using XGBoost, RF, ExtraTrees and softmax Regression as base learner classifiers. The outcome demonstrates the improved classification accuracy of the proposed approach for HAR compared to the modern methods and a simple and efficient training process for the model.

Nurhanim K et al. [25] demonstrated a smartphone's wearable sensors in HAR using Bagging, Adaboost, Rotation forest and Random subspace with RF as a base classifier. To determine the classification of each operation, the Holdout method and the ten-fold cross-validation method were performed. The results show that Adaboost-RF achieves 98.07% of holdout accuracy and 98.82% of 10 fold cross-validation methods.

3   Methodology

This section presents the methodology used in this study, including ensemble method and performance evaluations used in this study for Activity Recognition in ARAS dataset. The ARAS dataset from two real houses, House A and B were processed and assessed using python high-level programming language.

3.1   Ensemble Method

The study employed two ensemble classifiers, Bagging and Adaboost ensemble methods, by considering five popular classification algorithms. The study employed RL, RF, KNN, NB, and SVM as a base learner of ensemble classifiers to predict performance in ADLs in smart homes using ARAS dataset. These five classifiers were selected because of their representation in different types of algorithms with different feature representation and bias. The Bagging and Adaboost ensemble methods estimate different base models and use majority voting to combine the individual forecasts to achieve the final prediction. The estimates for each label are summed, and the label with the majority vote is predicted. These ensemble methods and five base learner classifiers have been used in this study because are simple to implement, faster, give better results, improve the model prediction's robustness, and overcome overfitting compared to other methods [26].

3.2   Data Preprocessing and Feature Selection

Data pre-processing was performed to prepared data for model training. Several methods were applied, such as one-hot encoding to convert the ARAS dataset into a form that classifiers can train a robust model that leads to better prediction. After that, feature selection techniques were employed to remove irrelevant features in the ARAS dataset to reduce computing complexity models. Feature Selection was also employed to reduce overfitting, give better accuracy, enable faster training, and reduce the model difficulty. The filter methods were used as it is preferred to be best than other methods such as embedded and wrapper methods based on their performance measure, irrespective of the classifier employed. Therefore, in this study, filtering techniques were used to prepare our data before modelling the data [27]. The ARAS dataset was then split into training data (80%) and testing (20%), training data was used to build the HAR model, and testing data was used to evaluate the performance of the HAR Model. The ARAS dataset is the imbalanced dataset to solve this problem, and this work employed the undersampling technique because it deals with majority class > 10K observations. After that, two ensemble methods were applied together with five base learner classifiers were to build HAR Model. The performance evaluation is conducted to evaluate model performance. Figure 1 below shows the suggested method for HAR in a smart home using an ensemble method and ARAS dataset carried out in this research.

4   Experimental Results and Discussions

This section presents the experiment results and discussion for the suggested method for HAR in multiclass classification ARAS dataset collected from two houses, House A and B, in the smart homes environment.

4.1   Experimental Dataset

In this experiment, the ARAS dataset was used to carter for recognition of ADL in smart homes. ARAS was collected from two different real houses for two months. It consists of 27 other activities, as per Table 1 below. Activities collected every day were as follows: shower, toiletries, breakfast preparations, breakfast, dinner, dinner, sleep, snacks, television and learning, reading books; while activities not performed every day, on the other hand, were like washing dishes, napping, laundering, shaving, telephone chat, etc. The total number of occurrences of events in House A and House B is 2177 and1023, respectively. There were 86400 data points for each day, consisting of the time stamp, and the prediction on both House A and B shows in this study that residents frequently performed going out, sleeping, studying, watching TV and getting breakfast [28, 29]. The dataset was split into training and testing set by taking 80% and 20% of the dataset, and the training set was used to develop the predictive models. In contrast, the testing set was used to assess the models' efficiency using both selected ensemble methods.

The ARAS dataset is an imbalanced dataset with an unequal number of instances in each class, making it more challenging to analyze accurate activity recognition. In this regard, the metrics performance was changed to provide better insight; this included precision, recall, and f1-score.

Secondly, we tried different classifiers to increase the performance; this helped improve the performance for both the f1 score and recall, although the accuracy score was slightly lower. RF performed well with the imbalanced dataset compared to other algorithms. Thirdly, we applied Resampling Techniques; under resample technique, there are two techniques; undersampling technique and oversample technique. The undersampling technique caters for the majority class, while oversampling method deals with the minority class. In this approach, The study used the undersampling technique, which attempts to balance class distribution by randomly removing the majority class distribution. The study employed the undersampling technique instead of the oversampling technique because our dataset contains the majority class >10K observations. In comparison, the oversampling approach works best with the dataset with <10K observations. To randomly delete samples from the majority class, the resampling module was then applied from Scikit-Learn. In this case, the majority of the class was dropped, resulting in a small dataset of the ratio between the 1:1. After that, five base learner classifiers were applied one by one, and RF outperformed compare to other classifiers as shown in Table 2 and Fig. 2 and Table 3 and Fig. 3 in the results and discussions section.

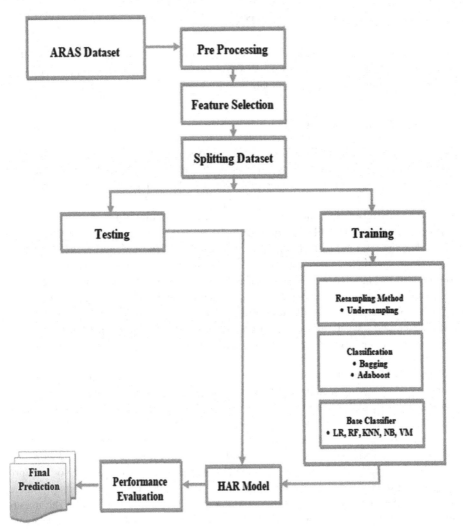

Fig. 1. Proposed Ensemble Approach

Table 1. List of Activities in ARAS Dataset

ID	Activity	ID	Activity	ID	Activity
1	Others	10	Having Snack	19	Laundry
2	Going Out	11	Sleeping	20	Shaving

(continued)

Table 1. (*continued*)

ID	Activity	ID	Activity	ID	Activity
3	Preparing Breakfast	12	Watching TV	21	Brushing Teeth
4	Having Breakfast	13	Studying	22	Talking on the Phone
5	Preparing Lunch	14	Having shower	23	Listening to Music
6	Having Lunch	15	Toileting	24	Cleaning
7	Preparing Dinner	16	Napping	25	Having Conversation
8	Having Dinner	17	Using Internet	26	Having Guest
9	Washing Dishes	18	Reading Book	27	Changing Clothes

4.2 Performance Evaluations

This research employed four performance evaluations to measure the performance of our model:

$$\text{Pr}ecision = \frac{(TP)}{(TP + FP)} \tag{1}$$

$$\text{Re}call\frac{(TP)}{(TP + FN)} \tag{2}$$

$$F1 - Score = 2*\frac{(\text{Pr}ecision * \text{Re}call)}{(\text{Pr}ecision + \text{Re}call)} \tag{3}$$

$$Accuracy = \frac{(TP + TN)}{(TP + TN + FP + FN)} \tag{4}$$

Where TP—True Positive, TN—True Negative, FP—False Positive, and FN—False Negative.

4.3 Results and Discussions

The experiment was done in two phases using two selected ensemble methods; Table 2 shows the Bagging method's results. Table 3 shows the results obtained using Adaboost ensemble methods from both House A and B in the ARAS dataset.

Table 2 presents the Bagging method's performance from five selected base learners together with precision, recall, and F1-Score from both House A and B. It was noted that the overall Random Forest (RF) outperformed in both House A and B compared to other classifiers, in House A, obtained an average accuracy of 96.2% and 98.4% in House B based on the Bagging Ensemble Method. Also, the results in Table 2 are displayed in Fig. 2 below.

Table.2. Overall assessment of Bagging Method

Base Classifiers	House A				House B			
	Acc (%)	Precision (%)	Recall (%)	F1-Score (%)	Acc (%)	Precision (%)	Recall (%)	F1-Score (%)
RL	0.90	0.90	0.91	0.90	0.92	0.92	0.92	0.92
RF	0.96	0.96	0.97	0.96	0.98	0.98	0.97	0.98
KNN	0.62	0.57	0.60	0.61	0.76	0.75	0.77	0.76
NB	0.22	0.23	0.22	0.22	0.24	0.24	0.24	0.23
SVM	0.88	0.88	0.78	0.88	0.92	0.90	0.92	0.90

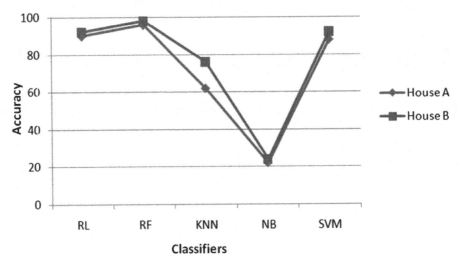

Fig.2. Overall assessment of Bagging Method

Table 3. Overall assessment of Adaboost Method

Base Classifiers	House A				House B			
	Acc (%)	Precision (%)	Recall (%)	F1-Score (%)	Acc (%)	Precision (%)	Recall (%)	F1-Score (%)
RL	0.92	0.92	0.91	0.92	0.94	0.94	0.92	0.94
RF	**0.97**	**0.96**	**0.97**	**0.97**	**0.99**	**0.99**	**0.98**	**0.99**
KNN	0.74	0.70	0.71	0.74	0.91	0.90	0.90	0.91
NB	0.25	0.39	0.24	0.26	0.30	0.33	0.32	0.30
SVM	0.93	0.93	0.94	0.93	0.95	0.95	0.95	0.96

As shown in Table 3 below, experimental was done using the Adaboost ensemble method with five selected base learners, including precision, recall, and F1-Score for both House A and B. The Random Forest (RF) outperformed in both House A and B with an accuracy of 97.1% and 99.2%, respectively, as a base learner compared to SVM at 93% and 95% in House A and B. Also, the results in Table 3 are displayed in Fig. 3 below.

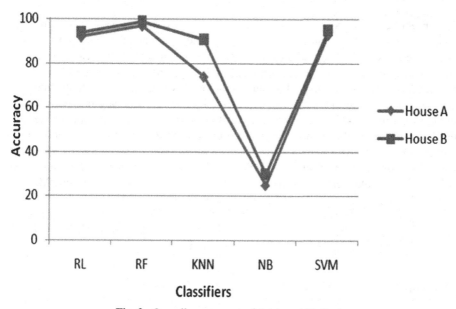

Fig. 3. Overall assessment of Adaboost Method

Table 4. Comparisons of Prediction Results on ARAS Dataset with previous research work

Research Study	Method	Accuracy -Average Score House A	Accuracy -Average Score House B
Jethanandani et al. [13]	Majority Voting Ensemble Classifier method	88%	97%
Alemdar, Hande, et al. [18]	Hidden Markov model (HMM)	61.5%	76.2%
This study	Adaboost Ensemble Classifier Method with RF	97.1%	99.2%

As shown in Table 4, the results show that the Adaboost Ensemble Method outperformed in ARAS Dataset based on RF classifier as learner compared to the previous studies. This study achieved an average score of 97.1% accuracy in House A and 99.2% accuracy in House B. In contrast to the earlier studies, which achieved 88% and 97% accuracy in House A and B respectively [13] and 61.5% in average score in House A and 76.2% accuracy for House B [18].

5 Conclusion and Future Work

The ensemble method's suggested technique has improved HAR performance in ADL predictions using ARAS dataset in smart homes. The experimental result shows that the prediction results have been enhanced in ARAS dataset by performing feature selection and ensemble methods compared to previous studies done using other ensemble methods and single prediction methods. The study employed two ensemble methods with five base learners. The results show that RF produced a better accuracy rate with 96.2% in House A and 98.4% accuracy in House B compared to other classifiers based on the Bagging Ensemble Method. In comparison, the Adaboost ensemble method also outperformed with RF compared to other classifiers with the accuracy of 97.1% in House A and 99.2% of accuracy in House B. In overall performance evaluation, the Adaboost ensemble method outperformed compared to the Bagging ensemble Method while RF outperformed compared to other classifiers in both House A and B. This study suggests different ensemble methods for future work and various types of feature selection methods to be applied in the ARAS dataset to increase the model's efficiency for recognizing ADLs for multi-residents in smart homes.

References

1. Kwon, M.C., Choi, S.: Recognition of daily human activity using an artificial neural network and smartwatch. Wireless Communications and Mobile Computing (2018)
2. Ogbuabor, G., Robert, L.: Human activity recognition for healthcare using smartphones. In: Proceedings of the 2018 10th International Conference on Machine Learning and Computing, pp. 41–46 (2018)
3. Xu, S., Tang, Q., Jin, L., Pan, Z.: A cascade ensemble learning model for human activity recognition with smartphones. Sensors 10, 2307 (2019)
4. Bughin, J., Manyika, J., Woetzel, J.: Smart Cities: digital solutions for a more livable future. McKinsey Global Institute Belgium, WI, USA (2018)
5. Wu, Y., Ke, Y., Chen, Z., Liang, S., Zhao, H., Hong, H.: Application of alternating decision tree with AdaBoost and bagging ensembles for landslide susceptibility mapping. CATENA 187, 104396 (2020)
6. Rokni, S.A., Nourollahi, M., Ghasemzadeh, H.: Personalized human activity recognition using convolutional neural networks. In: Proceedings of the AAAI Conference on Artificial Intelligence, 32(1) (2018)
7. Kee, Y.J., Zainudin, M.S., Idris, M.I., Ramlee, R.H., Kamarudin, M.R.: Activity recognition on subject independent using machine learning. Cybern. Info. Technol. 20(3), 64–74 (2020)
8. Abidine, B.M., Fergani, L., Fergani, B., Oussalah, M.: The joint use of sequence features combination and modified weighted SVM for improving daily activity recognition. Pattern Anal. Appl. 21(1), 119–138 (2016). https://doi.org/10.1007/s10044-016-0570-y

9. Prossegger, M., Bouchachia, A.: Multi-resident activity recognition using incremental decision trees. In: International Conference on Adaptive and Intelligent Systems, pp. 182–191. Springer, Cham (2014)

10. Elamvazuthi, I., Izhar, L.I., Capi, G.: Classification of human daily activities using ensemble methods based on smartphone inertial sensors. Sensors **12**, 4132 (2018)

11. Adama, D.A., Lotfi, A., Langensiepen, C., Lee, K., Trindade, P.: Human activity learning for assistive robotics using a classifier ensemble. Soft. Comput. **22**(21), 7027–7039 (2018). https://doi.org/10.1007/s00500-018-3364-x

12. Altuve, M., Lizarazo, P., Villamizar, J.: Human activity recognition using improved complete ensemble EMD with adaptive noise and long short-term memory neural networks. Biocybernetics Biomed. Eng. 40(3), 901–909 (2020)

13. Padmaja, B., Prasad, V.R., Sunitha, K.V.N.: A novel random split point procedure using extremely randomized (Extra) trees ensemble method for human activity recognition. EAI Endorsed Trans. Pervasive Health Technol. **6**(22), e5 (2020)

14. Chen, W.H., Chen, Y.: An ensemble approach to activity recognition based on binary sensor readings. In: 2017 IEEE 19th International Conference on e-Health Networking, Applications and Services (Healthcom), IEEE, pp. 1–5 (2017)

15. Chen, Z., Jiang, C., Xie, L.: A novel ensemble ELM for human activity recognition using smartphone sensors. IEEE Trans. Industr. Inf. **15**(5), 2691–2699 (2018)

16. Irvine, N., Nugent, C., Zhang, S., Wang, H., Ng, W.W.: Neural network ensembles for sensor-based human activity recognition within smart environments. Sensors **1**, 216 (2020)

17. Jethanandani, M., Sharma, A., Perumal, T., Chang, J.R.: Multi-label classification based ensemble learning for human activity recognition in smart home. Internet of Things **12**, 100324 (2020)

18. Ni, Q., Zhang, L., Li, L.: A heterogeneous ensemble approach for activity recognition with the integration of change point-based data segmentation. Appl. Sci. **8**(9), 1695 (2018)

19. Saha, J., Chowdhury, C., Roy Chowdhury, I., Biswas, S., Aslam, N.: An ensemble of condition-based classifiers for device-independent detailed human activity recognition using smartphones. Information **9**(4), 94 (2018)

20. Bulbul, E., Cetin, A., Dogru, I.A.: Human activity recognition using smartphones. In: 2018 2nd international symposium on multidisciplinary studies and innovative technologies (ismsit), IEEE, pp. 1–6 (2018)

21. Subasi, A., et al.: Sensor based human activity recognition using adaboost ensemble classifier. Procedia Comput. Sci. **140**, 104–111 (2018)

22. Tian, Y., Wang, X., Chen, W., Liu, Z., Li, L.: Adaptive multiple classifiers fusion for inertial sensor-based human activity recognition. Clust. Comput. **22**(4), 8141–8154 (2019)

23. Tian, Y., Wang, X., Chen, L., Liu, Z.: Wearable sensor-based human activity recognition via two-layer diversity-enhanced multiclassifier recognition method. Sensors **19**(9), 2039 (2019)

24. Xu, S., Tang, Q., Jin, L., Pan, Z.: A cascade ensemble learning model for human activity recognition with smartphones. Sensors **19**(10), 2307 (2019)

25. Nurhanim, K., Elamvazuthi, I., Izhar, L.I.: Ensemble Methods for Classifying of Human Activity Recognition. In: 2018 IEEE 4th International Symposium in Robotics and Manufacturing Automation (ROMA), IEEE, pp. 1–5 (2018)

26. Hartmann, J., Huppertz, J., Schamp, C., Heitmann, M.: Comparing automated text classification methods. Int. J. Res. Market. **36**(1), 20–38 (2019)

27. Brownlee, J.: Machine learning mastery with Python: understand your data, create accurate models, and work projects end-to-end. Mach. Learn. Mastery **527**, 100–120 (2016)

28. Alemdar, H., Ertan, H., Incel, O.D., Ersoy, C.: ARAS human activity datasets in multiple homes with multiple residents. In: 2013 7th International Conference on Pervasive Computing Technologies for Healthcare and Workshops, IEEE, pp. 232–235 (2013)
29. Alemdar, H., Ersoy, C.: Multi-resident activity tracking and recognition in smart environments. J. Ambient. Intell. Humaniz. Comput. **8**(4), 513–529 (2017). https://doi.org/10.1007/s12652-016-0440-x

Correction to: Application of Ensemble Techniques Based Sentiment Analysis to Assess the Adoption Rate of E-Learning During Covid-19 Among the Spectrum of Learners

S. Sirajudeen⑩, Balaganesh⑩, Haleema⑩, and V. Ajantha Devi⑩

Correction to:
Chapter "Application of Ensemble Techniques Based
Sentiment Analysis to Assess the Adoption Rate of E-Learning
During Covid-19 Among the Spectrum of Learners"
in: A. Solanki et al. (Eds.): *Artificial Intelligence*
***and Sustainable Computing for Smart City*, CCIS 1434,**
https://doi.org/10.1007/978-3-030-82322-1_14

In the originally published version of chapter 14 the affiliation of the first author was incorrect. The affiliation has been corrected as "Lincoln University College, Petaling Jaya, Malaysia".

The updated version of this chapter can be found at
https://doi.org/10.1007/978-3-030-82322-1_14

Author Index

Printed in the United States
by Baker & Taylor Publisher Services